Table of Contents

Introduction

This books intention is to cover in detail, theoretically and practically, everything needed for Monte Carlo and Quantum Monte Carlo methods, with its applications and basic examples. Since it is written as a guide, it does not contain the applications to the complex systems investigated today, but it comes right at their doorstep. Thus, it may be considered as beginners and intermediate level guide.

The problem the author observed with other books on the subject is that they under- or over-describe the method, thus the reader is either left confused or loses precious time. This book started as a guide for postgraduate students, and it continued in that fashion – to be clear and understandable, while covering all the material. At the end of the book is the bibliography list – it can be said that this book is the union of all the important parts from all of those books listed, with added code, images and explanations where the authors did not describe the problem properly.

The book starts with a short introduction to random number generators and statistics, progresses to the Monte Carlo method in general, Markov Chain, Markov Chain Monte Carlo, and finally - Quantum Monte Carlo section. It touches on the subject of the Hidden Monte Carlo.

As for the Quantum Monte Carlo, Variational Monte Carlo and Diffusion Monte Carlo methods are described in detail, including problems like fixed-node approximation, importance sampling, Metropolis algorithm, etc.

Program code is in C++, and it relies heavily on Numerical Recipes library. However, the algorithms are clearly presented and the code is understandable and commented so it can be ported to other programming languages.

Acknowledgements

Some parts of the code are from Numerical Recipes book. The reader is advised to obtain Numerical Recipes libraries for either C++ or Java. If the reader wishes not to follow algorithms only, but to build on the code given in this book, then the NR libraries are a necessity. As it is the ultimate guide, the NR book is responsible for most of the references in the non-Quantum sections. However, it is a difficult book to follow, because it is very dense with information, and the step-by-step guide is omitted. For further understanding of numerical computation, the NR book is the best refence.

I wish to thank prof. PhD. Goran Poparic, from Physics College, University of Belgrade for guiding me through this book. The book will be used as a textbook on the postgraduate class prof. Poparic holds: "Monte Carlo simulations in Physics".

Required knowledge

1. Any programming language, preferable C++,
2. Some basic background in statistics,
3. For Quantum Monte Carlo section, the required is some knowledge in Quantum physics and chemistry,
4. Desirable is some knowledge in Solid State physics and molecular physics.

Abbreviations

- PRNG – pseudo-random number generator
- # - number
- Def – definition
- Max, min – maximum, minimum
- NR – Numerical Recipes book
- MC – Monte Carlo
- MCMC – Markov Chain Monte Carlo
- HMM – Hidden Markov Model
- iff – if and only if
- QMC – Quantum Monte Carlo
- VMC – Variational Monte Carlo
- DMC – Diffusion Monte Carlo
- PIMC – Path Integral Monte Carlo
- CLT – Central Limit Theorem
- Δ. – Proof
- SE – Schrodinger equation
- TISE – Time independent SE
- HF – Hartree-Fock
- LHO – linear harmonic oscillator

1. Generators of pseudo-random numbers[1]

Def. PRNG is an algorithm for generating a sequence of numbers whose properties approximate the properties of random numbers.

PRNG generated sequence is not truly random - pseudo-random are computer generated; random is from nature.

Initial state – seed state – it will always produce the same sequence when initialized with that state.

The period: the max of the length of the repetition-free prefix of the sequence. If internal state contains n bits, its period can be no longer than 2^n results.

PRNG's can be split up to two groups:

1. Uniform generators – produce sequences which are uniformly distributed (a finite number of values are equally likely to be observed; every one of n values has equal probability $1/n$).
2. Non-uniform generators (pseudo-random number sampling or non-uniform pseudo-random variate generation) – produce sequences which are distributed according to a given probability distribution. These sorts of deviates are almost always generated by performing appropriate operations on one or more uniform deviates.

1.1 Uniform deviates

They lie between 0.0 and 1.0 for floating-point numbers or between 0 and $2^{32} - 1$ or $2^{64} - 1$ for integers.

1.1.1 Recommended methods for use in Combined Generators

We like algorithms with few and fast operations, with foolproof initialization and with state small enough to keep in registers or first-level cache. To be recommendable for use in combined generator, the method must be understood theoretically and it must pass empirical tests. The period, the set of returned values, and the set of valid initializations should be completely understood.

As minimal empirical standard, tests that can be used:

- Diehard battery of statistical tests (that includes "Gorilla test"),
- NIST-STS.

It is important to require that each method in a combined generator separately pass the tests → the combination generator should be better than any single component.

Abbreviations:

- * - means that a method passes the Diehard test by itself (for 64b quantities 32 high and low bits each pass).
- "Can be used as random" – minimal level for applications where a combined generator is not needed.

---

[1] This section is only a quick overview of the PRNG and it is not crucial to know how they work in order to use them. NR suggests using **Ran** and **Ranq** generators in **ran.h**, and for using PRNG in MC that is all we need to know. Because of that, some of the things are not explained in detail, but there is enough of information online for anyone who wishes to investigate further.

- "Can improve by" - the recommended combination.
- "Can use as a rand" – HQ,
- "Can use in bit mix" – LQ, unless they pass Diehard.

Methods that use 64b unsigned arithmetic: `Ullong` – unsigned long long (Linux) – unsigned `__int64` (MS)

64b XOR-shift method

In just three XORs and three shifts it produces a full period of $2^{64} - 1$ on 64b. Zero must be avoided. Generator can use the rule that starts with << or >>[2].

- state: x (unsigned 64b)
- initialize: $x \neq 0$
- update: $x \leftarrow x \wedge (x \gg a_1), x \leftarrow x \wedge (x \ll a_2), x \leftarrow x \wedge (x \gg a_3)$
- can use as random: x (all bits) *
- can use in bit mix: x (all bits)
- can improve by: output 64b MLCG successor
- period: $2^{64} - 1$

XOR is the same as addition → all three lines in update can be written as the action of a 64×64 matrix on a vector, where matrix is all zeroes except for ones on the diagonal and on exactly one super- (<<) or sub-diagonal (>>) - S_k ($k > 0$ for left-shift, $k < 0$ for right-shift) → One full step of updating: $T = S_{k_3} S_{k_2} S_{k_1}$.

One needs to find triples of integers (k_1, k_2, k_3) that give the full $M = 2^{64} - 1$ period. Necessary and sufficient conditions are: $T^M = 1$ and that $T^N \neq 1$ for $N = M/\{6700417, 655537, 641, 257, 17, 5, 3\}$.

Only a small fraction of full-period triples produce generators that pass Diehard.

ID	a_1	a_2	a_3
A1	21	35	4
A2	20	41	5
A3	17	31	8
A4	11	29	14
A5	14	29	11
A6	30	35	13
A7	21	37	4
A8	21	43	4
A9	23	41	18

---

[2] Bitwise shift

Multiply with Carry (MWC) with Base $b = 2^{32}$

The base is chosen to be a power of 2 that is half the available word length, $b = 32$ for 64b words. MWC is defined by its multiplier a.

- state: x (unsigned 64b)
- initialize: $1 \leq x \leq 2^{32} - 1$
- update: $x \leftarrow a \, (x \, \& \, [2^{32} - 1]) + (x \gg 32)$
- can use as random: x (low 32b) *
- can use in bit mix: x (all 64b)
- can improve by: output 64b XOR-shift successor to 64b x
- period: $(2^{32}a - 2)/2$, a – prime

Although only the low b bits of state x can be taken as random, there is considerable randomness in all the bits of x that represent the product ab, which is convenient in a combined generator.

ID	a
B1	4294957665
B2	4294963023
B3	4162943475
B4	3947008974
B5	3874257210
B6	2936881968
B7	2811536238
B8	2654432763
B9	1640531364

CB Modulo 2^{64}

Its high 32b almost pass Diehard, its low 32b are a disaster, but this generator has its place in the construction of combined generators.

- state: x (unsigned 64b)
- initialize: any value
- update: $x \leftarrow ax + c \pmod{2^{64}}$
- can use as rand: x (high 32b, with caution)
- can use in bit mix: x (high 32b)
- can improve by: output 64b XOR-shift successor
- period: 2^{64}

ID	a	c (any odd value ok)
C1	3935559000370003845	2691343689449507681
C2	3202034522624059733	4354685564936845319
C3	2862933555777941757	7046029254386353087

1.1.2 Constructing Combined Generators

The methods being combined:

i.   should be independent of one another,

ii.   they must share no state (although their initializations are allowed to derive from some convenient common seed),

iii.   should have different periods,

iv.   should look like each other algorithmically as little as possible,

v.   the output of the combination generator should not perturb the independent evolution of individual methods,

vi.   should be combined by binary operations whose output is no less random than one input if the other is fixed. For 32b and 64b unsigned only + and $\wedge$ operators can be used,

vii.   all bit positions in combined output should depend on high-quality bits from at least two methods, and may depend on lower-quality bits from additional methods.

Composition

Another trick is using a method as a successor relation instead of as own generator. Each of the methods above is mapping from 64b state x_i to successor state x_{i+1}. A good method must have no correlation between a state and its successor. If it has a period of 2^{64} or $2^{64} - 1$, then all values occur exactly once.

Labeling: if we take output of a generator C1 and run it through A6 we denote that as A6(C1) – composed generator – which has the period of C1.

Random mapping of C1's output fixes problem with low bits, and A6(C1) will fix A6's weakness that a bit depends only on a few bits of the previous state.

Direct combining

Is better than composition because it:

i.   increases the size of the state,

ii.   increases the length of the period.

iii.   It does not increase the size of the state or length of the period.

Composition is superior to direct combining because it mixes widely differing bit positions. E.g. A6+C1 is not acceptable because low bits of C1 are poor, but A6(C1) has no such liability.

Ran()

Recommended Uniform generator is in `ran.h`, `struct Ran`. `Ran()` takes a single integer argument, which becomes the seed. An instance of Ran offers several different formats for rand output.

Example of instance by declaration:

```
Ran myran(17); //myran is instance, 17 is seed
myran.int64() // returns a rand 64b unsigned int
myran.int32() // returns a rand 32b unsigned int
myran.doub() //returns a double-precision floating value in the range 0.0 and 1.0
1 + myran.int64() % (n - 1) // returns a rand int between 1 and n
```

`Ran` is a combination of composition and direct combining with + and $\wedge$. It is the generator that is combination and composition of four different generators.

l – left-shift operation is done first, r – right-shift operation is done first.

$$Ran = [A1_l(C3) + A3_r] \wedge B1$$

The period is at least multiple of the periods of C3, A3 and B1.

Ranq()

It is in `ran.h`, named `Ranq1` for shorter, and `Ranq2` for longer period.

The simplest and fastest generator that is recommended:

$$Ranq1 = D1(A1_r)$$

It generates 64b random integer in 3 shifts, 3 XORs and one multiply; or a double floating value in one additional multiply. Its method is small enough to go inline. It should not be used by an application that makes more than 10^{12} calls, since its period is $1.8 \cdot 10^{19}$. Ranq is fine for 99.99% of applications and Ran can be used for the remaining 0.01%.

1.2 Methods for quality check of PRNG

Good statistical properties are a central requirement for the output of a PRNG. In general, careful mathematical analysis is required to have any confidence that a PRNG generates numbers that are sufficiently close to random to suit the intended use.

Program that produces random numbers should be uncorrelated with the one using them – any two different random number generators, coupled with applications program, should give the same statistical result. If they don't, then at least one is not a good generator.

PRNG which do not satisfy statistical pattern tests:

- shorter than expected periods for some seed states,
- lack of uniformity of distribution (uniform distribution – symmetric probability distribution where a finite # of values are equally likely to be observed),
- correlation of successive values,
- poor dimensional distribution,
- the distances between certain values are distributed differently from those truly random.

Never use:

- A generator principally based on a linear congruential generator (LCG) or a multiplicative linear congruential generator (MLCG).
- A generator with a period less than $\sim 2^{64} \approx 2 \cdot 10^{19}$, or any generator whose period is undisclosed.
- A generator that warns against using its low-order bits as being completely random. That was a good advice once, but now it indicates an obsolete algorithm (usually a LCG).
- The built-in generators in the C and C++ languages, especially `rand` and `srand`. These have no standard implementation and are often badly flawed.

Avoid using generators:

- That take more than 24 arithmetic or logical operations to generate a 64b integer or double precision floating result.
- Designed for cryptographic use.
- With period $> 10^{100}$ – we will not need it and the period of a generator has little to do with its quality.

An acceptable rand generator must combine at least two (ideally, unrelated) methods. The methods combined should evolve independently and share no state. The combination should be by simple operations that do not produce results less random than their operands.

1.2.1 Combined generators

Same as above, we like algorithms with few and fast operations, with foolproof initialization and with state small enough to keep in registers or first-level cache. To be recommendable for use in combined generator, the method must be understood theoretically and it must pass empirical tests. The period, the set of returned values, and the set of valid initializations should be completely understood.

As minimal empirical standard, tests that can be used:

- Diehard battery of statistical tests (that includes "Gorilla test"),
- NIST-STS.

It is important to require that each method in a combined generator separately pass the tests → the combination generator should be better than any of its components.

1.2.2 BSI evaluation criteria

BSI has established 4 criteria for quality of pseudo-random generators (the latter two are of cryptographic importance):

1. A sequence of random numbers with a low probability of containing identical consecutive elements.
2. A sequence of numbers which is indistinguishable from true random according to specified statistical tests:
   a. Monobit test – equal numbers of 1 and 0 in the sequence,
   b. Poker test – special instance of the chi-squared test,
   c. Runs test – counts the frequency of runs of various lengths – after a sequence of n zeros (or 1s), the next bit is one (or zero) with probability ½,
   d. Longruns test,
   e. Autocorrelation test.

2. Distributions, mean, standard deviation and dispersion[3]

2.1 Terminology

- $p(x)$ – probability density function (PDF) – density of a continuous random variable – function that describes the relative likelihood for this random variable to take on a given value. The probability of the variable falling within a range of values is given by the integral of this variable's density over that range – P – area under the density function. Integral over the entire space is equal to 1.

- Number of values – N. We count them: $x_0, \dots, x_{N-1}$, but if their mean is known a priori rather than being estimated we take: $x_1, \dots, x_N$.

- Median – x_{med} - 50% chance that it will fall to the left, 50% to the right, i.e. the value such that the set of values less than the median has a probability of ½: $\int_{-\infty}^{x_{med}} p(x)dx = \frac{1}{2} = \int_{x_{med}}^{\infty} p(x)dx$

- Mean – expected (central) value; for discrete the mean is the sum over every possible value weighted by the probability of that value: $\mu = \sum xP(x)$; for continuous: $\mu = \int_{-\infty}^{\infty} xf(x)dx$. If weight is the same for every value and there are N values $x_0, \dots, x_{N-1}$: $\bar{x} = \frac{1}{N}\sum_{j=0}^{N-1} x_j$

- Standard error: if we calculate $\bar{x}$ many times with different sets of N sampled data, the values $\bar{x}$ will have a standard deviation – standard error of $\bar{x}$. For Normal distribution, standard error of:
  - $\bar{x}: \frac{\sigma}{\sqrt{N}}$
  - $Var: \sigma^2 \sqrt{\frac{2}{N}}$
  - $\sigma: \frac{\sigma}{\sqrt{2N}}$

- Mode – maximum, the value with highest probability. Bimodal – distribution has two relative maxima. Example:

```
In[16]:= Plot[PDF[(NormalDistribution[0, 1.5]), x] + PDF[(NormalDistribution[5, 1.5]), x],
         {x, -10, 10}, Filling → Axis]
```

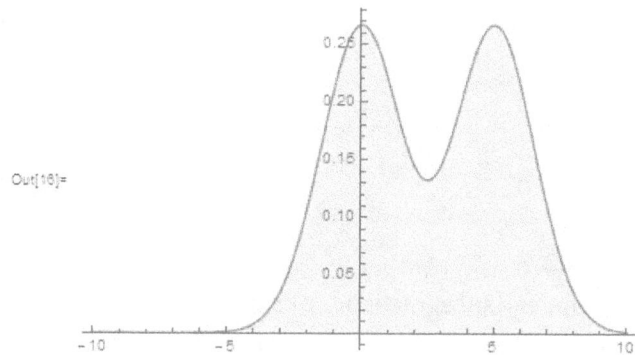

- $cdf = P(< x)$ – cumulative density function: $cdf = \int_{-\infty}^{x} p(x')\,dx'$

- Inverse cdf – Quantile function: given $p(x)$ (for example, $p(x) = 1 - e^{-\lambda Q}$) and percentile p, the quantile function is derived by finding the value for which $1 - e^{-\lambda Q} = p \rightarrow Q = -\frac{\ln(1-p)}{\lambda}$. Quantile is used in Monte Carlo methods. It is one way of prescribing $p(x)$ and is an alternative to $p(x)$, cdf and the characteristic function. MC employ quantile function to produce non-uniform

---

[3] The code given is all from the NR libraries.

random or pseudorandom numbers for use in diverse types of simulation calculations. A sample from a given distribution may be obtained, in principle, by applying its quantile function to a sample from a uniform distribution.

$x(P)$ – inverse of cdf – needed for finding the values of x associated with the specific percentile points or quantiles in significance tests.

- $\sim$ - "is drawn from a distribution"
- Skewness – a measure of the extent to which $p(x)$ leans to one side (from its mean) – a measure of asymmetry.

$$Skew(x_0, \ldots, x_{N-1}) = \frac{1}{N} \sum_{j=0}^{N-1} \left(\frac{x_j - \bar{x}}{\sigma} \right)^3$$

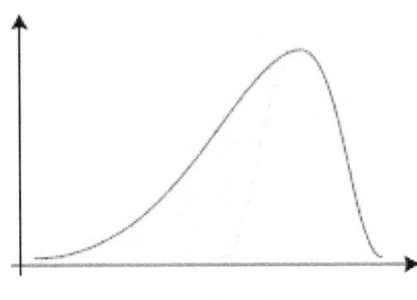 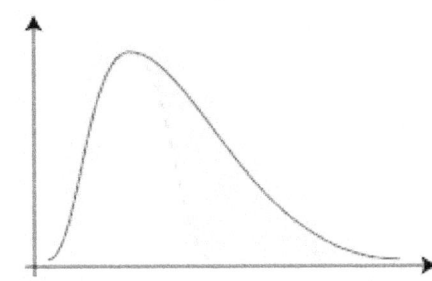

Negative Skew   Positive Skew

- Kurtosis: measures the relative peakness of a distribution relative to normal distribution.

$$Kurt(x_0, \ldots, x_{N-1}) = \left(\frac{1}{N} \sum_{j=0}^{N-1} \left(\frac{x_j - \bar{x}}{\sigma} \right)^4 \right) - 3$$

$$Kurt(Normal) = 0$$

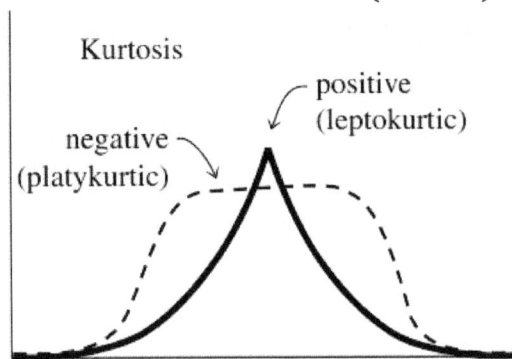

- σ - standard deviation – square root of the variance, measure of dispersion.
- $\sigma^2 \equiv Var$ - variance – the second moment of $p(x)$ about the mean, measure of the dispersion – width around the central value: $Var(x_0, \ldots, x_{N-1}) = \frac{1}{N-1} \sum_{j=1}^{N-1} (x_j - \bar{x})^2$.
- Tail – the large set of values where $p(x)$ is low,
- Head – the range of values where $p(x)$ is high.
- Moment – quantitative measure of the shape of a set of points.
  $\langle x^n \rangle = \int x^n P_X(x) dx$ – n^{th} moment.
  - 0th moment – area under the curve – equal to 1,
  - 1st moment – mean,
  - 2nd moment – variance,

- ○ 3rd moment – skewness,
- ○ 4th moment - kurtosis.

Code:

- Distribution's parameters are set by the constructor and then referenced as needed by the member functions.
- `p()` – density function.
- `cdf()` – cdf
- `invcdf()` – inverse cdf.

2.2 Normal (Gaussian) distribution

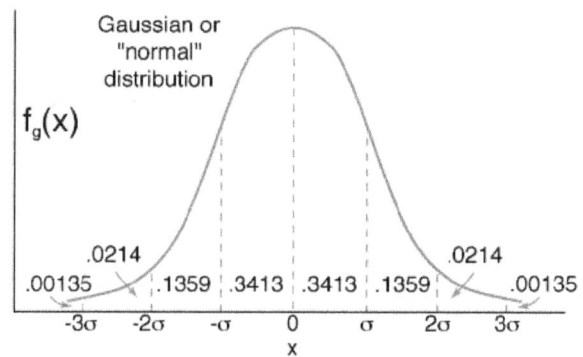

For large # of events Gauss distribution approximates Binomial distribution.

$$x \sim N(\mu, \sigma)$$

$$p(x) = \frac{1}{\sqrt{2\pi}\sigma} e^{-\frac{1}{2}\left(\frac{x-\mu}{\sigma}\right)^2}$$

$$cdf = \frac{1}{2} erfc\left(-\frac{1}{\sqrt{2}}\frac{x-\mu}{\sigma}\right)$$

$$x(P) = \mu - \sqrt{2}\sigma\, erfc^{-1}(2P)$$

$\mu \equiv \bar{x}$ – mean, σ – standard deviation, $\sigma^2 = \overline{x^2} - \bar{x}^2$ – variance

It is in `erf.h` with `struct Normaldist`.

Mean of a function $f(x)$: $\langle f(x)\rangle = \int f(x)p(x)dx$

$\langle x^n \rangle = \int x^n p(x)dx$ – nth moment of distribution.

2.3 Cauchy (Lorentzian) Distribution

Like Normal, Cauchy distribution is centrally peaked, symmetric with μ - center and σ – width, but it has tails that decay very slowly at $\pm\infty$ as $|x|^{-2}$ → only 0th moment exists.

$$x \sim Cauchy(\mu, \sigma)$$

$$p(x) = \frac{1}{\pi\sigma}\frac{1}{1 + \left(\frac{x-\mu}{\sigma}\right)^2}$$

$$x \sim Cauchy(0,1) \Rightarrow \frac{1}{x} \sim Cauchy(0,1) \Rightarrow \frac{1}{ax+b} \sim Cauchy\left(-\frac{b}{a}, \frac{1}{a}\right)$$

14

$$cdf = \frac{1}{2} + \frac{1}{\pi}\arctan\left(\frac{x-\mu}{\sigma}\right)$$

$$x(P) = \mu + \sigma\tan\left(\pi\left(P - \frac{1}{2}\right)\right)$$

μ is not mean, σ is not standard deviation.

It is in `distributions.h`, with `struct Cauchydist`.

2.4 Student-t distribution

A generalization of the Cauchy distribution. It has ν parameter that specifies how rapidly it decays: $|t|^{-(\nu+1)}$.

$$\nu \in \mathbb{N} \Rightarrow \nu = \text{number of convergent moments}$$

$$t \sim \text{Student}(\nu, \mu, \sigma), \nu > 0$$

$$p(t) = \frac{\Gamma\left(\frac{1}{2}(\nu+1)\right)}{\Gamma\left(\frac{1}{2}\nu\right)\sqrt{\nu\pi}\sigma}\left(1 + \frac{1}{\nu}\left(\frac{t-\mu}{\sigma}\right)^2\right)^{-\frac{1}{2}(\nu+1)}$$

$$\nu = 1 - \text{Cauchy}, \nu \to \infty - \text{Normal}$$

μ is mean, but σ^2 is not variance.

$$\text{Var}\{\text{Student}(\nu,\mu,\sigma)\} = \frac{\nu}{\nu-2}\sigma^2$$

$$cdf = \begin{cases} \frac{1}{2}I_x\left(\frac{1}{2}\nu,\frac{1}{2}\right), t \le \mu \\ 1 - \frac{1}{2}I_x\left(\frac{1}{2}\nu,\frac{1}{2}\right), t > \mu \end{cases}$$

I_x – inverse incomplete beta function.

Above form of distribution is rarely used, since most statistical tests using it are double-sided. The two-tailed function $A(t|\nu)$ is defined only for the case $\mu = 0$ and $\sigma = 1$:

$$A(t|\nu) = \int_{-t}^{t} p(t')dt' = 1 - I_x\left(\frac{1}{2}\nu,\frac{1}{2}\right)$$

A is notably used in the test of whether two observed distributions have the same mean.

It is in `incgammabeta.h` with `struct Studenttids`.

2.5 Logistic Distribution

Symmetric, centrally peak distribution that can be used instead of Normal. Its tails decay exp but slower than Normal distribution.

$$p(y) = \frac{e^{\pm y}}{(1+e^{\pm y})^2} = \frac{1}{4}\text{sech}^2\left(\frac{1}{2}y\right)$$

To avoid overflows, use negative form for y>0, positive for y<0.

$$\text{Var}\{\text{Logistic}\} = \frac{\pi^2}{3}$$

We use standardized logistic distribution because its μ is mean, and σ is the standard deviation.

$$x \sim \text{Logistic}(\mu, \sigma)$$

$$p(x) = \frac{\pi}{4\sqrt{3}\sigma} \operatorname{sech}^2\left(\frac{\pi}{2\sqrt{3}} \frac{x-\mu}{\sigma}\right)$$

$$cdf = \frac{1}{1 + e^{-\frac{\pi}{\sqrt{3}}\frac{x-\mu}{\sigma}}}$$

$$x(P) = \mu + \frac{\sqrt{3}}{\pi}\sigma \log\frac{P}{1-P}$$

It is in `distributions.h` with `struct Logisticdist`.

2.6 Exponential distribution

Defined on the $x \geq 0$, parameter β controls its width.

$$x \sim \text{Exponential}(\beta), \beta > 0$$

$$p(x) = \beta e^{-\beta x}$$

$$cdf = 1 - e^{-\beta x}$$

$$x(P) = -\frac{1}{\beta}\log(1-P)$$

$$\mu = \sigma = \frac{1}{\beta}$$

$$\text{Median} = \frac{\log 2}{\beta}$$

It is in `distributions.h` under `struct Expondist`.

2.7 Weibull distribution

Generalizes exponential distribution in a way useful in hazard, survival or reliability studies. When the lifetime (time to failure) of an item is exponentially distributed, there is a constant probability per unit time that an item will fail, if it has not already done so:

$$\text{hazard} = \frac{p(x)}{P(> x)} = \left(\frac{\alpha}{\beta}\right)\left(\frac{x}{\beta}\right)^{\alpha-1}, \alpha > 0$$

$\alpha > 1$ – hazard increases with time, $\alpha < 1$ – hazard decreases with time, $\alpha = 1$ – exponential distribution.

$$x \sim \text{Weibull}(\alpha, \beta) \text{ iff } y = \left(\frac{x}{\beta}\right)^{\alpha} \sim \text{Exponential}(1)$$

$$p(x) = \left(\frac{\alpha}{\beta}\right)\left(\frac{x}{\beta}\right)^{\alpha-1} e^{-\left(\frac{x}{\beta}\right)^\alpha}, x > 0$$

$$cdf = 1 - e^{-\left(\frac{x}{\beta}\right)^\alpha}$$

$$x(P) = \beta(-\log(1-P))^{\frac{1}{\alpha}}$$

Max is at: $x = \beta\left(\frac{\alpha-1}{\alpha}\right)^{\frac{1}{\alpha}}$; $\mu = \beta\Gamma(1+\alpha^{-1})$; $\sigma^2 = \beta^2\left(\Gamma(1+2\alpha^{-1}) - (\Gamma(1+\alpha^{-1}))^2\right)$

2.8 Lognormal distribution

Many processes that live on the positive x-axis are approximated by normal distributions on the $\log x$-axis, e.g. multiplicative random walk which starts at some positive value x_0, and then generates new values by a recurrence, like:

$$x_{i+1} = \begin{cases} x_i(1+\epsilon), p = .5 \\ x_i/(1+\epsilon), p = .5 \end{cases}$$

$$x \sim \text{Lognormal}(\mu, \sigma) \text{ iff } u = \frac{\log(x) - \mu}{\sigma} \sim N(0,1)$$

$$p(x) = \frac{1}{\sqrt{2\pi}\sigma x} e^{-\frac{1}{2}\left(\frac{\log x - \mu}{\sigma}\right)^2}$$

While μ and σ are the mean and the standard deviation in $\log x$-space, they are not in x-space:

$$\text{Mean}\{\text{Lognormal}(\mu,\sigma)\} = e^{\mu+\frac{1}{2}\sigma^2}, \text{Var}\{\text{Lognormal}(\mu,\sigma)\} = e^{2\mu+\sigma^2}(e^{\sigma^2} - 1)$$

$$cdf = \frac{1}{2}erfc\left(-\frac{1}{\sqrt{2}}\frac{\log x - \mu}{\sigma}\right)$$

$$x(P) = e^{\mu-\sqrt{2}\sigma erfc^{-1}(2P)}$$

It is in `erf.h` under `struct Lognormaldist`.

2.9 Chi-Square distribution

χ^2 distribution has a single parameter $\nu > 0$ which controls location and width of distribution's peak. $\nu \in \mathbb{N}$ and is # of degrees of freedom.

$$\chi^2 \sim \text{Chisquare}(\nu)$$

$$p(\chi^2)d\chi^2 = \frac{1}{2^{\frac{1}{2}\nu}\Gamma\left(\frac{1}{2}\nu\right)}(\chi^2)^{\frac{1}{2}\nu-1}e^{-\frac{1}{2}\chi^2}d\chi^2$$

Mean is ν, variance is equal to 2ν. It is a special case of the Gamma distribution.

$$cdf = P\left(\frac{\nu}{2}, \frac{\chi^2}{2}\right) = 1 - Q\left(\frac{\nu}{2}, \frac{\chi^2}{2}\right)$$

$$x(P) = 2P^{-1}\left(\frac{\nu}{2}, P\right)$$

It is in `incgammabeta.h` under `struct Chisqdist`.

2.10 Gamma distribution

$$x \sim \text{Gamma}(\alpha, \beta), \alpha > 0, \beta > 0$$

$$p(x) = \frac{\beta^\alpha}{\Gamma(\alpha)} x^{\alpha-1} e^{-\beta x}, x > 0$$

Exponential distribution is the special case with $\alpha = 1$. χ^2 distribution is the special case, with $\alpha = \frac{\nu}{2}$ and $\beta = \frac{1}{2}$.

$$\text{Mean}\{\text{Gamma}(\alpha, \beta)\} = \frac{\alpha}{\beta}, \text{Var}\{\text{Gamma}(\alpha, \beta)\} = \frac{\alpha}{\beta^2}$$

$$cdf = P(\alpha, \beta x)$$

$$x(P) = \frac{1}{\beta} P^{-1}(\alpha, P)$$

It is in `incgammabeta.h` under `struct Gammadist`.

2.11 F-distribution

Tests whether two observed samples have the same variance.

Parameters: d_1, d_2 – degrees of freedom.

$$f = \frac{\sqrt{(d_1, x)^{d_1} d_2^{d_2}}}{xB\left(\frac{d_1}{2}, \frac{d_2}{2}\right)}$$

2.12 Beta distribution

$$x \sim \text{Beta}(\alpha, \beta), \alpha > 0, \beta > 0$$

$$p(x) = \frac{1}{B(\alpha, \beta)} x^{\alpha-1} (1 - x)^{\beta-1}, x \in (0,1), B - \text{beta function}$$

$$\text{Mean}\{\text{Beta}(\alpha, \beta)\} = \frac{\alpha}{\alpha + \beta}$$

$$\text{Var}\{\text{Beta}(\alpha, \beta)\} = \frac{\alpha\beta}{(\alpha + \beta)^2 (\alpha + \beta + 1)}$$

$\alpha > 1 \land \beta > 1$ – there is a single mode at $(\alpha - 1)(\alpha + \beta - 2)$; when $\alpha < 1 \land \beta < 1$ distribution function is U-shaped with a min at this same value. In other cases there is neither a max nor a min.

$$cdf = I_x(\alpha, \beta)$$

$$x(P) = I_p^{-1}(\alpha, \beta)$$

It is in `incgammabeta.h` under `struct Betadist`.

2.13 Kolmogorov-Smirnov KS distribution
Used to find if two distributions are different.

2.14 Poisson distribution
It is a border line case of binomial distribution when $n \gg k \Rightarrow p \ll$.

Binomial distribution is: $p(k) = \binom{n}{k} p^k (1-p)^{n-k}$

Approximations:

- $\binom{n}{k} = \frac{n!}{k!(n-k)!} = \frac{n(n-1)\cdots(n-k+1)}{k!} \to \frac{n^k}{k!}, n \to \infty$

- $(1-p)^{n-k} \approx (1-p)^n = (1-p)^{\frac{np}{p}} = |np = \lambda = \text{const.}| = (1-p)^{\frac{\lambda}{p}} \approx \left| \lim_{p \to 0} (1-p)^{\frac{1}{p}} = e^{-1} \right| \approx e^{-\lambda}$

It applies to a process where discrete, uncorrelated events occur at some mean rate per unit time. If, for a given period, λ is the mean expected # of events → probability distribution of seeing k events:

$$k \sim \text{Poisson}(\lambda)$$

$$p(k) = \underbrace{n^k p^k}_{\lambda^k} \frac{e^{-\lambda}}{k!} = \frac{1}{k!} \lambda^k e^{-\lambda}, k \in \mathbb{N}_0$$

$$\sum_k p(k) = 1$$

$\langle k \rangle = \sum_k k p(k) = \lambda$ - Mean and Var are λ. There is a single mode at $k = \lfloor \lambda \rfloor$.

$cdf = P_\lambda(< k) = Q(k, \lambda)$, Q – incomplete gamma function

$$P_\lambda(< 0) = 0, P_\lambda(< 1) = e^{-\lambda}, P_\lambda(< \infty) = 1$$

It is in `incgammabeta.h` under `struct Poissondist`.

2.15 Binomial distribution
We observe n-particle gas in a volume V, and label a part of that volume as v. We are interested in probability of finding k molecules inside v. If we could numerate those n molecules, the probability that those are the first k molecules is: $p(k) = p^k (1-p)^k$.

It is the discrete distribution over $k \geq 0$. It has two parameters, $n \geq 1$ – sample size or max value for which k can be nonzero; and p – event probability (not to be confused with $p(k)$ – the probability of a particular k).

$$k \sim \text{Binomial}(n, p), n \geq 1, p \in (0,1)$$

$$p(k) = \underbrace{\binom{n}{k}}_{\text{\# of permutations}} \underbrace{p^k}_{\substack{\text{probability of event} \\ \text{happening n-times}}} \underbrace{(1-p)^{n-k}}_{\substack{\text{probability that something} \\ \text{else will happen}}} , k \in \mathbb{N}_0, \binom{n}{k} = \frac{n!}{k!\,(n-k)!}$$

$p(k)$ – the probability of finding random k molecules from n total molecules in v.

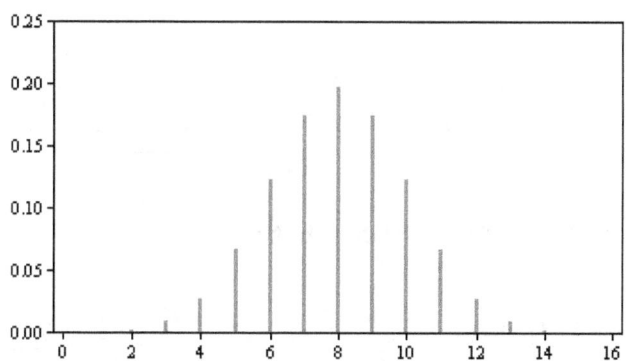

$$\text{Mean}\{\text{Binomial}(n,p)\} = np, \text{Var}\{\text{Binomial}(n,p)\} = np(1-p)$$

There is a single mode at k that satisfies: $(n+1)p - 1 < k \le (n+1)p$

$p = \frac{1}{2} \rightarrow$ distribution is symmetrical, otherwise it has positive skewness for $p < \frac{1}{2}$ and negative for $p > \frac{1}{2}$.

$n \to \infty, p \to 0, np < \infty \rightarrow$ Poisson distribution.

$$cdf = P(< k) = 1 - I_p(k, n-k+1)$$

It is in `incgammabeta.h` under `struct Binomialdist`.

Example: Random movement in 1D discrete space

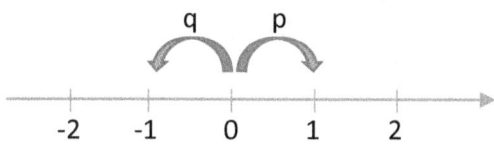

p – probability of going right, q – left.

N – total # of steps, n_1 - # of steps to the right, n_2 – to the left.

$$N = n_1 + n_2, m = n_1 - n_2$$

Probability of finding after 2 steps:

$$\left(pe^{i\varphi} + qe^{-i\varphi}\right)^2 = p^2 e^{2i\varphi} + 2pq + q^2 e^{-2i\varphi} \Rightarrow \begin{cases} p^2, \text{ probability of finding at 2} \\ q^2, \text{ at -2} \\ 2pq, \text{ at 0} \end{cases}$$

Probability after N steps: $\left(pe^{i\varphi} + qe^{-i\varphi}\right)^N$

Probability that after N steps it is at m:

$$P_N(m) = \frac{1}{2\pi} \int_{-\pi}^{\pi} \left(pe^{i\varphi} + qe^{-i\varphi}\right)^N e^{-im\varphi} d\varphi = \frac{1}{2\pi} \int_{-\pi}^{\pi} \sum_{k=0}^{N} \binom{N}{k} q^{N-k} p^k e^{-i\varphi(N-k)} e^{-im\varphi} d\varphi$$

$$= \frac{1}{2\pi} \sum_{k=0}^{N} \binom{N}{k} \int_{-\pi}^{\pi} p^k q^{N-k} e^{i(2k-N-m)} d\varphi$$

Bernoulli distribution (e.g. Binary $p = 1, q = 0$; e.g. coin toss):
probabilities are normalized $\rightarrow p = 1 - q$

$$\frac{1}{2\pi}\int_{-\pi}^{\pi} e^{i\varphi s}d\varphi = \delta(s,0)$$

$$P_N(m) = \sum_{k=0}^{N}\binom{N}{k}p^k(1-p)^{N-k}\underbrace{\delta(2k-N-m)}_{\delta\left(k,\frac{N+m}{2}\right)} = \underbrace{\binom{N}{\frac{N+m}{2}}}_{\frac{N!}{\left(\frac{1}{2}(N+m)\right)!\left(\frac{1}{2}(N-m)\right)!}} p^{\frac{N+m}{2}}(1-p)^{\overbrace{\frac{N-m}{2}}^{N-\frac{N+m}{2}}} =$$

$$\frac{N!}{\left(\frac{1}{2}(N+m)\right)!\left(\frac{1}{2}(N-m)\right)!}p^{\frac{N+m}{2}}(1-p)^{\frac{N-m}{2}} - \textbf{Binomial distribution}$$

Symmetric case: $(p = q = \frac{1}{2})$: $P_N(m) = \dfrac{N!}{\left(\frac{N+m}{2}\right)!\underbrace{\left(N-\frac{N+m}{2}\right)}_{\frac{N-m}{2}}!}\underbrace{\dfrac{1}{2^{\frac{N+m}{2}}}\dfrac{1}{2^{\frac{N-m}{2}}}}_{2^{-N}} = 2^{-N}\dfrac{N!}{\left(\frac{1}{2}(N+m)\right)!\left(\frac{1}{2}(N-m)\right)!}$

3. Modeling of various non-uniform generators: Graphical and Analytical method

"Pseudo-random number sampling" or "non-uniform pseudo-random variate generation" is a numerical practice of generating pseudo-random numbers that are distributed according to a given probability distribution. Methods of sampling a non-uniform distribution are typically based on the availability of a pseudo-random number generator, producing numbers x that are uniformly distributed. Computational algorithms are then used to manipulate a single random variate, x, or often several such variates, into a new random variate y such that these values have the required distribution.

Random deviates with uniform probability between 0 and 1, denoted $U(0,1)$, have probability:

$$p(x)dx = \begin{cases} dx, x \in [0,1) \\ 0, \text{ otherwise} \end{cases}, x{\sim}U(0,1)$$

What we need: given invertible $p(y) = f(y)$, we need to find $x = F(y) = cdf = P(< y) = \int_0^y p(y)dy$, which will give us the desired $y(x) = F^{-1}(x): p(y) \rightarrow F(y) \rightarrow y(x)$

In[32]:= Integrate[PDF[(NormalDistribution[3, 1.5]), x] + PDF[(NormalDistribution[7, 1.5]), x], {x, 0, x}]

Out[32]= 0.977248 - 0.5 Erf[1.41421 - 0.471405 x] - 0.5 Erf[3.29983 - 0.471405 x]

In[36]:= Plot[{PDF[(NormalDistribution[3, 1.5]), x] + PDF[(NormalDistribution[7, 1.5]), x], Out[32] / 4}, {x, -5, 10}, Filling → Axis]

Out[36]=

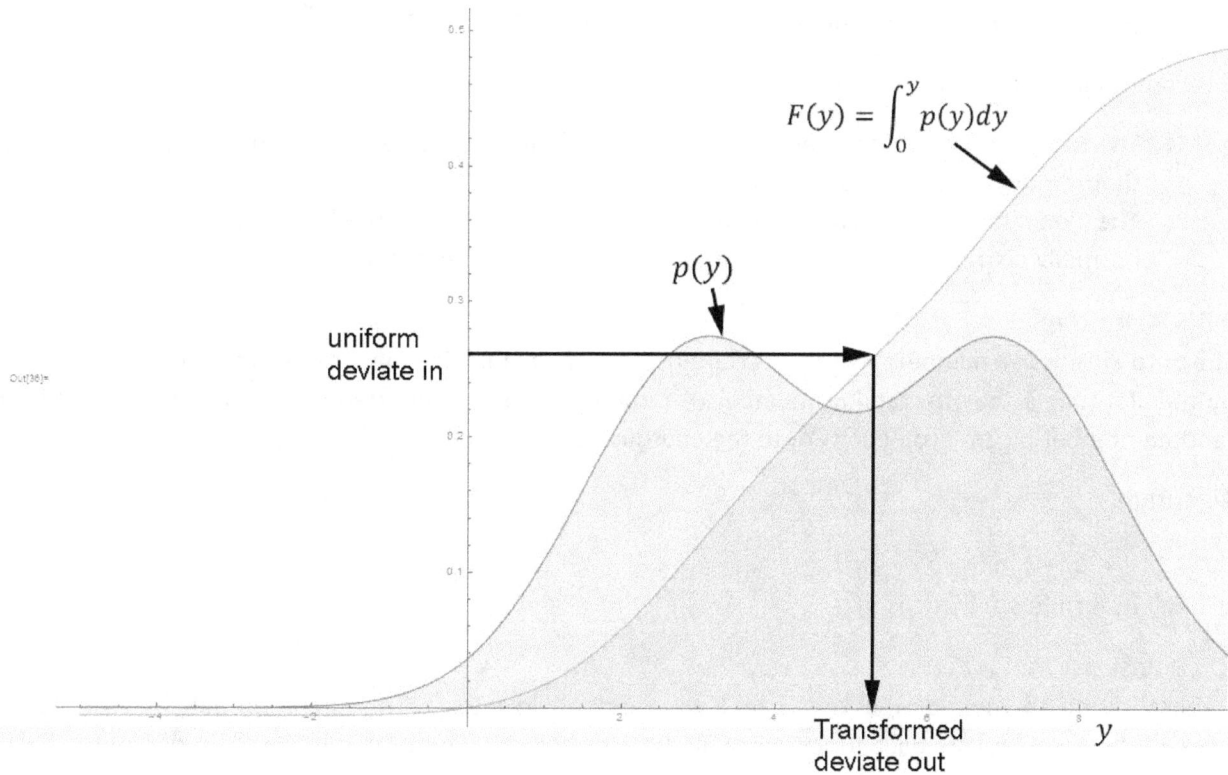

3.1 Exponential deviates

$$p(y) = \beta e^{-\beta y} \rightarrow F(y) = \int_0^y \beta e^{-\beta y} dy = 1 - e^{-\beta y} \rightarrow y(x) = -\frac{1}{\beta}\ln(1-x)$$

It is in `deviates.h` under `struct Expondev`.

3.2 Logistic deviates

$$p(y) = \frac{\pi}{4\sqrt{3}\sigma}\operatorname{sech}^2\left(\frac{\pi}{2\sqrt{3}}\frac{y-\mu}{\sigma}\right) \rightarrow F(y) = \frac{1}{1+e^{-\frac{\pi}{\sqrt{3}}\frac{y-\mu}{\sigma}}} \rightarrow x = y + \frac{\sqrt{3}}{\pi}\sigma\log\frac{F(y)}{1-F(y)}$$

It is in `deviates.h` in struct `Logisticdev`.

3.3 Normal deviates by transformation (Box-Muller transform)

The Box-Muller transform is a pseudo-random number sampling method for generating pairs of independent, standard, normally distributed (zero expectation, unit variance) random numbers, given a source of uniformly distributed random numbers. It produces a pair of Gaussian random numbers from a pair of uniform numbers. It utilizes the fact that a 2D distribution of two independent zero-mean Gaussian random numbers is radially symmetric if both component Gaussians have the same variance. This can be seen by multiplying two 1D distributions: $e^{-x^2}e^{-y^2} = e^{-(x^2+y^2)} = e^{-r^2}$.

It is commonly expressed in two forms:

1. The Basic form: takes two samples from the uniform distribution on the interval $(0,1]$ and maps them to two standard, normally distributed samples.
2. The Polar form: takes two samples from a different interval $[-1,1]$ and maps them to two normally distributed samples without the use of trigonometry functions.

3.3.1 Algorithm

The output Gaussian numbers represent coordinates on 2D plane. The magnitude of the vector is obtained by transforming a uniform random number. A random phase is generated by scaling a second uniform random number by 2π. Projections onto the coordinate axes give Gaussian numbers.

Algorithm:

1. $a \leftarrow \sqrt{-2\ln U_1}, b \leftarrow 2\pi U_2$
2. Return $(a\sin b, a\cos b)$

3.3.2 Basic form

Generalization to multi-dimension: if $\{x_i\}$ are random deviates with a joint probability distribution $p(\{x_i\})d\{x_i\}$ and $\{y_i\}$ are each functions of all the x's, then the joint probability distribution of the y's is:

$$p(\{y_i\})d\{y_i\} = p(\{x_i\})\left|\frac{\partial\{x_i\}}{\partial\{y_i\}}\right|d\{y_i\}$$

Consider the transformation between two uniform deviates on $(0,1)$:

$$\left.\begin{cases} y_1 = \sqrt{-2\ln x_1}\cos 2\pi x_2 \\ y_2 = \sqrt{-2\ln x_1}\sin 2\pi x_2 \end{cases}\right\} \Rightarrow \begin{cases} x_1 = e^{-\frac{1}{2}(y_1^2 + y_2^2)} \\ x_2 = \dfrac{1}{2\pi}\arctan\dfrac{y_2}{y_1} \end{cases}$$

Jacobian determinant is:

$$\left|\frac{\partial(x_1,x_2)}{\partial(y_1,y_2)}\right| = \begin{vmatrix} \dfrac{\partial x_1}{\partial y_1} & \dfrac{\partial x_1}{\partial y_2} \\ \dfrac{\partial x_2}{\partial y_1} & \dfrac{\partial x_2}{\partial y_2} \end{vmatrix} = -\left(\frac{1}{\sqrt{2\pi}}e^{-\frac{y_1^2}{2}}\right)\left(\frac{1}{\sqrt{2\pi}}e^{-\frac{y_2^2}{2}}\right)$$

Since this is the product of a function of y_2 alone and the y_1 alone, we see that each y is independently distributed according to the normal distribution.

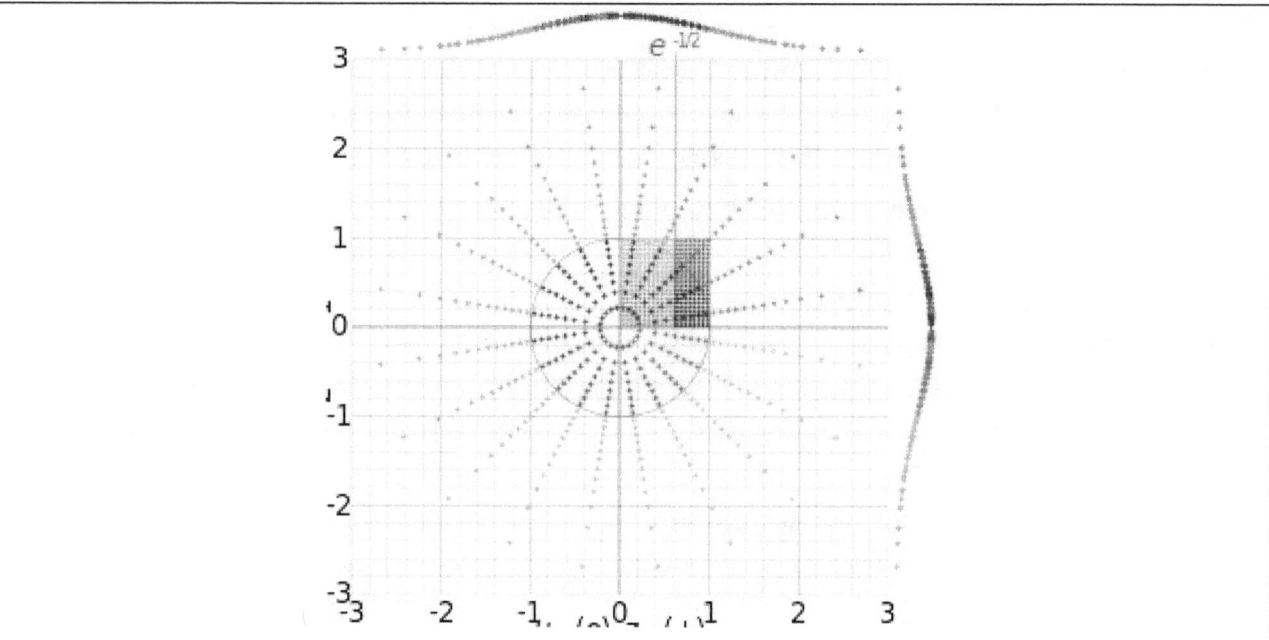

Visualization of the Box-Muller transform – the colored points in the unit square (x_1, x_2) are mapped to a 2D Gaussian (y_1, y_2). The plots at the top and the right are probability distribution functions of y_1 and y_2.

3.3.3 Polar form

In terms of calculating speed it is better to convert this to polar coordinates, since we will lose the expensive trigonometry functions.

Instead of x_1 and x_2 we pick v_1 and v_2 as the ordinate and abscissa of a random point inside the unit circle. Then $R^2 = v_1^2 + v_2^2$ is a uniform deviate, which can be used for x_1, while the angle which (v_1, v_2) defines with respect to the v_1-axis can serve as the random angle $\theta = 2\pi x_2$.

$$\begin{cases} \cos\theta = \dfrac{v_1}{R} \\ \sin\theta = \dfrac{v_2}{R} \end{cases} \Rightarrow \begin{cases} y_1 = \sqrt{-2\ln R^2}\,\dfrac{v_1}{R} = v_1\sqrt{-\dfrac{2\ln R^2}{R^2}} \\ y_2 = \sqrt{-2\ln R^2}\,\dfrac{v_2}{R} = v_2\sqrt{-\dfrac{2\ln R^2}{R^2}} \end{cases}$$

It is in `deviates.h` under `struct Normaldev_BM`.

3.4 Rejection method[4]

It is a powerful, general technique for generating random deviates whose distribution function $p(x)dx$ is known and computable.

Draw a graph of $p(x)$. On the same graph draw a comparison function $f(x)$ that has a finite area and lies above $p(x)$.

Imagine that we have some way of choosing a random point in 2D that is uniform in the area under the comparison function. Whenever that point lies outside the area under the original probability distribution, we will reject it and choose another random point. Whenever it lies inside, we will accept it.

The fraction of points rejected just depends on the ratio of the area of the comparison function to the area of the probability distribution function.

It remains only to suggest how to choose a uniform random point in 2D under the $f(x)$. Choose $f(x)$ whose indefinite integral is known analytically and also invertible to give x as a cdf. Pick a uniform deviate beween 0 and A (A is the total area under $f(x)$), and use it to get a corresponding x. Then pick a uniform deviate between 0 and $f(x)$ as the y value for the 2D point. Finally, accept or reject according to whether it is less or greater than $p(x)$.

To summarize, the rejection method requires that one finds a good comparison function. Each deviate requires two uniform random deviates, one evaluation of f to get y, and one evaluation of p to decide whether to accept or reject the point (x,y). The process is repeated, on the average, A times before the final deviate is obtained.

---

[4] In QMC this method will become useful as the part of the Importance Sampling.

```
In[15]:= Integrate[PDF[NormalDistribution[5, 1.5], x], {x, 0, x}]
Out[38]= 0.499571 - 0.5 Erf[2.35702 - 0.471405 x]

In[72]:= Plot[{PDF[(NormalDistribution[5, 1.5]), x], PDF[(NormalDistribution[5, 1.5]), x] (1 - Sin[2 x]^2/4)/1.1, Out[38]/2}, {x, 0, 10}, Filling → Axis]
```

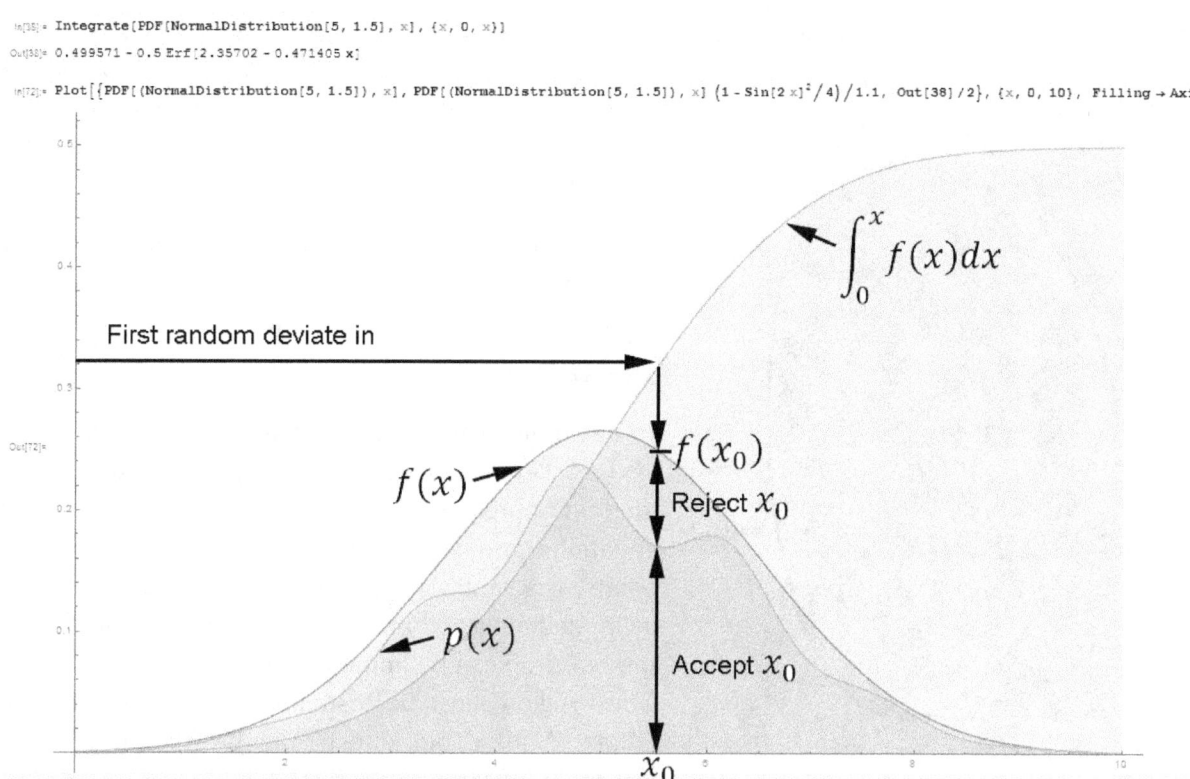

Rejection method for generating a random deviate x from a known probability distribution $p(x) \leq f(x)$. The transformation method is first used to generate a random deviate x of the distribution f. A second uniform deviate is used to decide whether to accept or reject x. If it is rejected, a new deviate of f is found, and so on. The ratio of accepted to rejected points is the ratio of the are under p to the area between p and f.

3.5 Ratio-of-Uniforms method

Typically ratio-of-uniforms method is used when the region can be closely bounded by a rectangle, parallelogram, or some other shape that is easy to sample uniformly. Then, we go from sampling the easy shape to sampling the desired region by rejection of points outside the desired region.

The ratio-of-uniforms has an advantage over the Box-Muller in that the square root is replaced by cheaper division, and that the logarithm function can be avoided in some cases. The disadvantage is that two uniform random numbers are consumed, but one Gaussian number produced.

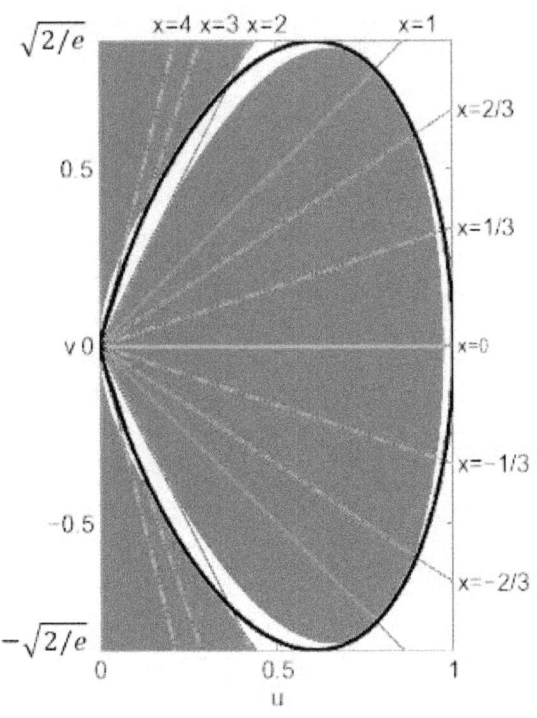

Each of the axes u and v correspond to one of the input uniform random numbers. Points enclosed by teardrop $|v| < \sqrt{-4u^2 \ln u}$. Teardrop equation: $v = \pm\sqrt{-4u^2 \ln u}$.

To avoid unnecessary evaluation of the exact boundary of the acceptance region, most implementations of this method approximate the region by using less complex equations to avoid computing the logarithm for many points. The central grey region contains points that can be immediately accepted, while the grey regions outside the teardrop can be immediately rejected. Points in the white region must be tested against the exact curve. Either of these quick tests can be eliminated if the logarithm function is very fast.

Algorithm:

1. loop
2. $u \leftarrow U_1$
3. $x \leftarrow \dfrac{V_1\sqrt{2/e}}{u}$
4. if $x^2 \leq 5 - 4e^{\frac{1}{4}}u + 1.4$ then //test for quick accept
5. return x
6. else if $x^2 < \dfrac{4e^{-1.35}}{u} + 1.4$ then //test for quick accept
7. if $v^2 < -4u^2 \ln u$ then //Do full test against exact curve
8. return x
9. end if
10. end if
11. end loop

3.5.1 Method by Leva

Tighter bounds are made with ellipse fitted to the curve. Deviates of any probability distribution $p(x)$ can be generated:

1. Construct the region in (u,v) plane bounded by $0 \leq u \leq \sqrt{p\left(\frac{v}{u}\right)}$,
2. Choose two deviates, u and v, that lie uniformly in the region,
3. Return v/u as the deviate.

$$s = .449871, t = -.386595, r_1 = .27597, a = .19600, b = .25472, r_2 = .27846$$

1. loop
2. $u \leftarrow U_1, v \leftarrow \sqrt{2/e}V_1$
3. $x \leftarrow u - s, y \leftarrow |v| - t$
4. $Q \leftarrow x^2 + y(ay - bx)$
5.   if $Q < r_1$ then return v/u
6.   else if $Q < r_2$ then
7.   if $v^2 < -4u^2 \ln u$ then return v/u
8. end loop

It is in `deviates.h` under `struct Normaldev`.

3.6 Gamma Deviates

For $\alpha > 1$ there is a rejection method based on a simple transformation of the gamma distribution combined with a squeeze. After transformation, the gamma distribution can be bounded by a Gaussian curve whose area is never more than 5% greater than that of the gamma curve.

It is in `deviates.h` under `struct Gammadev`.

3.7 Distributions from other deviates

From normal, gamma and uniform deviates many distributions are easily generated. We just need to make sure that when combining the results that all distinct instances of `Normaldist`, `Gammadist` and `Ran` have different random seeds.

3.7.1 Chi-Square Deviates

$$\chi^2(v) \cong \gamma\left(\frac{v}{2}, \frac{1}{2}\right) \cong 2\gamma\left(\frac{v}{2}, 1\right)$$

3.7.2 Student-t Deviates

It can be generated by a method similar to Box-Muller.

$$y = \sqrt{v\left(u_1^{-\frac{2}{v}} - 1\right)} \cos 2\pi u_2$$

We can use normal and gamma deviates.

$$x \sim N(0,1), y \sim \gamma\left(\frac{v}{2}, \frac{1}{2}\right) \Rightarrow x\sqrt{v}, y \sim \text{Student}(v, 0, 1)$$

3.7.3 Beta Deviates

$$x \sim \gamma(\alpha, 1), y \sim \gamma(\beta, 1) \Rightarrow \frac{x}{x + y} \sim \beta(\alpha, \beta)$$

3.8 Poisson Deviates

$$k \sim \text{Poisson}(\lambda)$$

$$p(k) = \frac{1}{k!} \lambda^k e^{-\lambda}, k \in \mathbb{N}_0$$

$$\sum_k p(k) = 1$$

$\langle k \rangle = \sum_k k p(k) = \lambda$ - Mean and Var are λ. There is a single mode at $k = \lfloor \lambda \rfloor$.

$cdf = P_\lambda(< k) = Q(k, \lambda)$, Q – incomplete gamma function.

Poisson distribution is a discrete distribution → deviates will be integers - k. It is convenient to convert the Poisson distribution into a continuous distribution by a trick: consider $p(k)$ as being spread uniformly into the interval $(k, k + 1)$ → this defines a continuous distribution by making k integer:

$$q_\lambda(k)dk = \frac{1}{\lfloor k \rfloor!} \lambda^{\lfloor k \rfloor} e^{-\lambda}$$

If we now use a rejection method to generate a deviate from this equation, and then take the integer part of that deviate, it will be as if drawn from the discrete Poisson distribution.

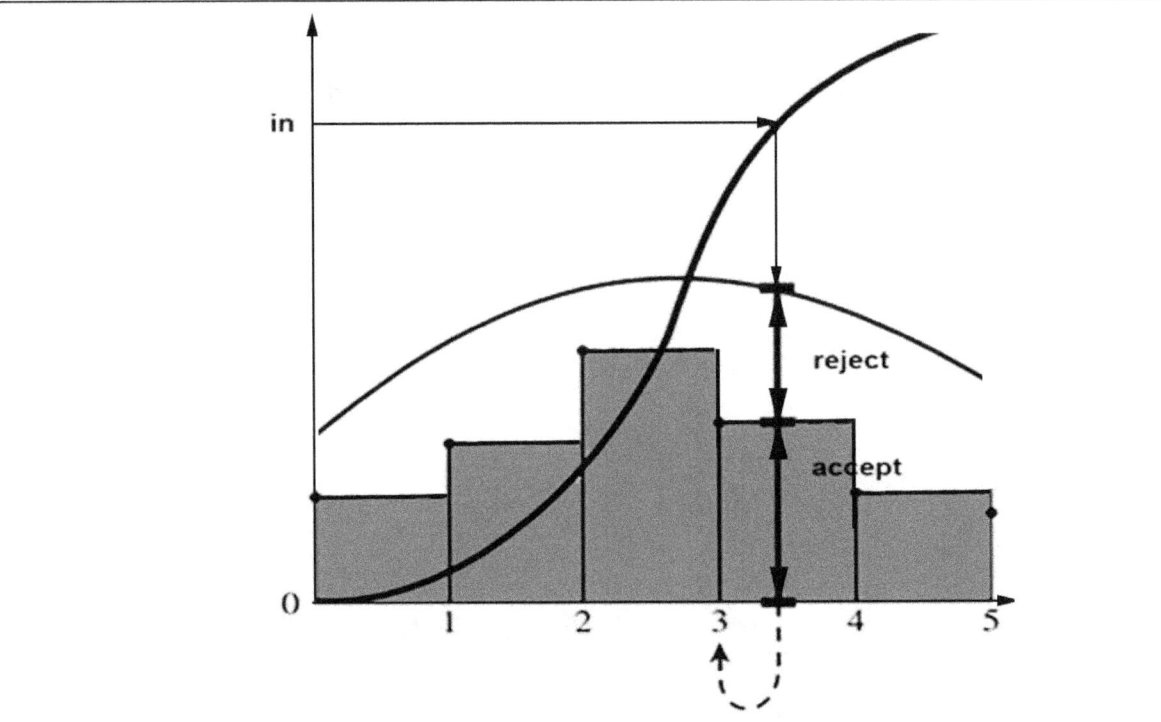

Rejection method as applied to an integer-valued distribution. The method is performed on the step function shown as a dashed line, yielding a real-valued deviate. This deviate is rounded down to the next lower integer, which is output.

This trick is general for any integer-valued probability distribution.

For λ large enough, the distribution is Bell-shaped → ratio-of-uniforms method works well. It is easy to find simple inner and outer squeezes in the (u, v) plane of the form $v^2 = Q(u)$, where $Q(u)$ is a simple polynomial.

It is in `deviates.h` under `struct Poissondev`.

3.9 Binomial Deviates

The distribution is integer valued → we use the same trick as with Poisson deviates to convert it into a stepped continuous distribution.

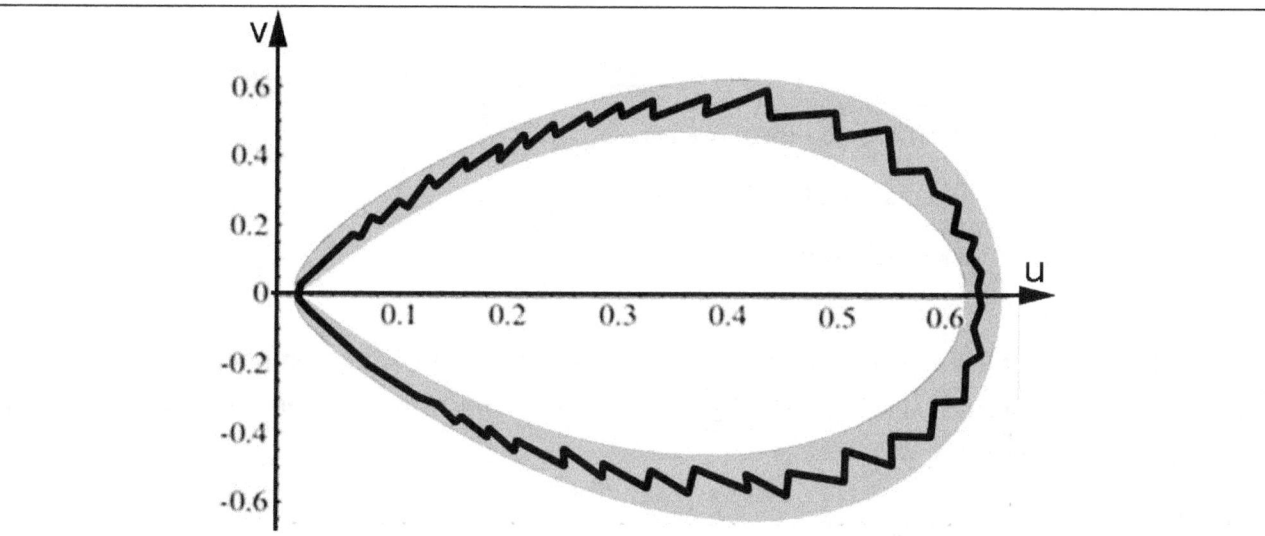

Ratio of uniforms method as applied to the generation of binomial deviates. Points are chosen randomly in the (u, v)-plane. The smooth curves are inner and outer squeezes. The jagged curves correspond to various binomial distributions with $n > 64$ and $np > 30$. An evaluation of the binomial probability is required only when the random point falls between the smooth curves.

Three cases:

1. $n > 64, np > 30$: use squeezed ratio-of-uniforms method,
2. $n > 64, np < 30$: use cdf lookup by bisection,
3. $n < 64$: use bit-parallel random comparison.

All three cases are in `deviates.h` under `struct Binomialdev`.

4. Error function and inverse Error function

The error function and complementary error function are special cases of the incomplete gamma function.

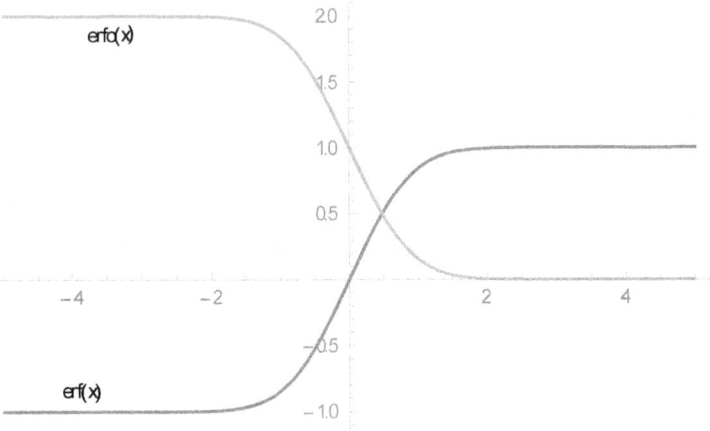

$$\text{erf}\, x = \frac{2}{\sqrt{\pi}} \int_0^x e^{-t^2}\, dt$$

$$\text{erfc}\, x = 1 - \text{erf}\, x = \frac{2}{\sqrt{\pi}} \int_x^{\infty} e^{-t^2}\, dt$$

Limiting values: $\text{erf}\, 0 = 0, \text{erf}\, \infty = 1, \text{erfc}\, 0 = 1, \text{erfc}\, \infty = 0$

Symmetries: $\text{erf}(-x) = -\text{erf}\, x, \text{erfc}\, 0 = 1, \text{erfc}(-x) = 2 - \text{erfc}\, x$

They are in **erf.h** under **struct Erf**.

4.1 Uses for Error Function

When the results of a series of measurements are described by a normal distribution with standard deviation σ and expected value $0 \rightarrow \text{erf}\left(\frac{a}{\sigma\sqrt{2}}\right)$ is the probability that the error of a single measurement lies within the interval $[-a, +a]$, for $a > 0$.

Linear correlation:

$(x_i, y_i), i = 0, \dots, N-1 \;\rightarrow\; r = \frac{\Sigma_i(x_i-\bar{x})(y_i-\bar{y})}{\sqrt{\Sigma_i(x_i-\bar{x})}\sqrt{\Sigma_i(y_i-\bar{y})}}$ — linear correlation coefficient (product-moment correlation coefficient or Pearson's r).

$r \in [-1,1]$:

- $r = 1$ - Complete positive correlation – data points lie on a straight line with positive slope, with x and y increasing together.

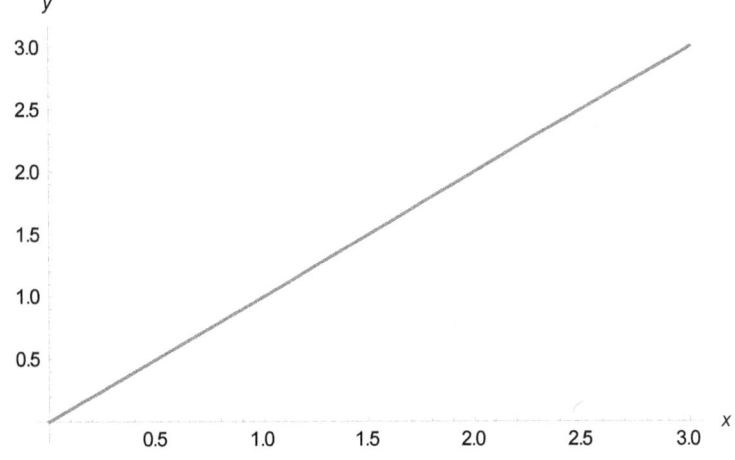

- $r = -1$ – Complete negative correlation – negative slope.

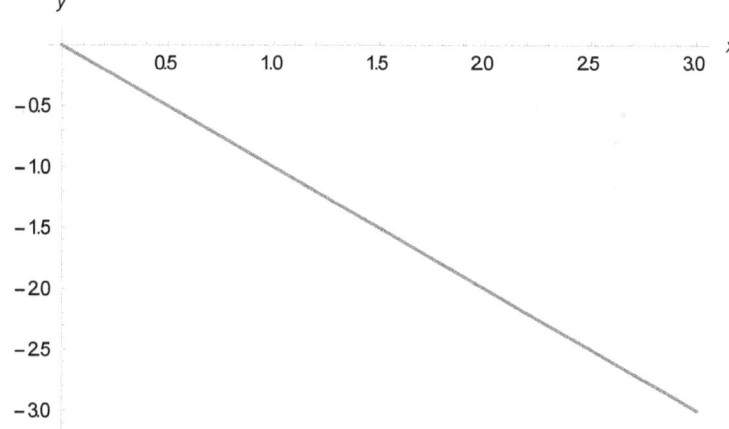

- $r \approx 0$ – uncorrelated.

r is poor statistic for deciding whether correlation is statistically significant and/or whether one observed correlation is significantly stronger than another.

If the null hypothesis is that x and y are uncorrelated $\rightarrow r$ is distributed approximately normally, with a mean of 0 and $\sigma = \frac{1}{\sqrt{N}}$ $\rightarrow$ significance of the correlation – the probability that $|r|$ should be larger than its observed value in the null hypothesis is: $\text{erfc}\frac{|r|\sqrt{N}}{\sqrt{2}}$ – if it is small, the two distributions are significantly correlated.

5. Central Limit Theorem CLT[5]

1. Draw n values from any distribution: $x_0, \dots, x_{n-1}$,
2. Take a mean of those n values: $\bar{x} = \sum_{j=0}^{n-1} x_j$,
3. Create a histogram of those $\bar{x}$ and observe that,
4. When $n \to \infty \to$ histogram of $\bar{x}$ will become a normal distribution.

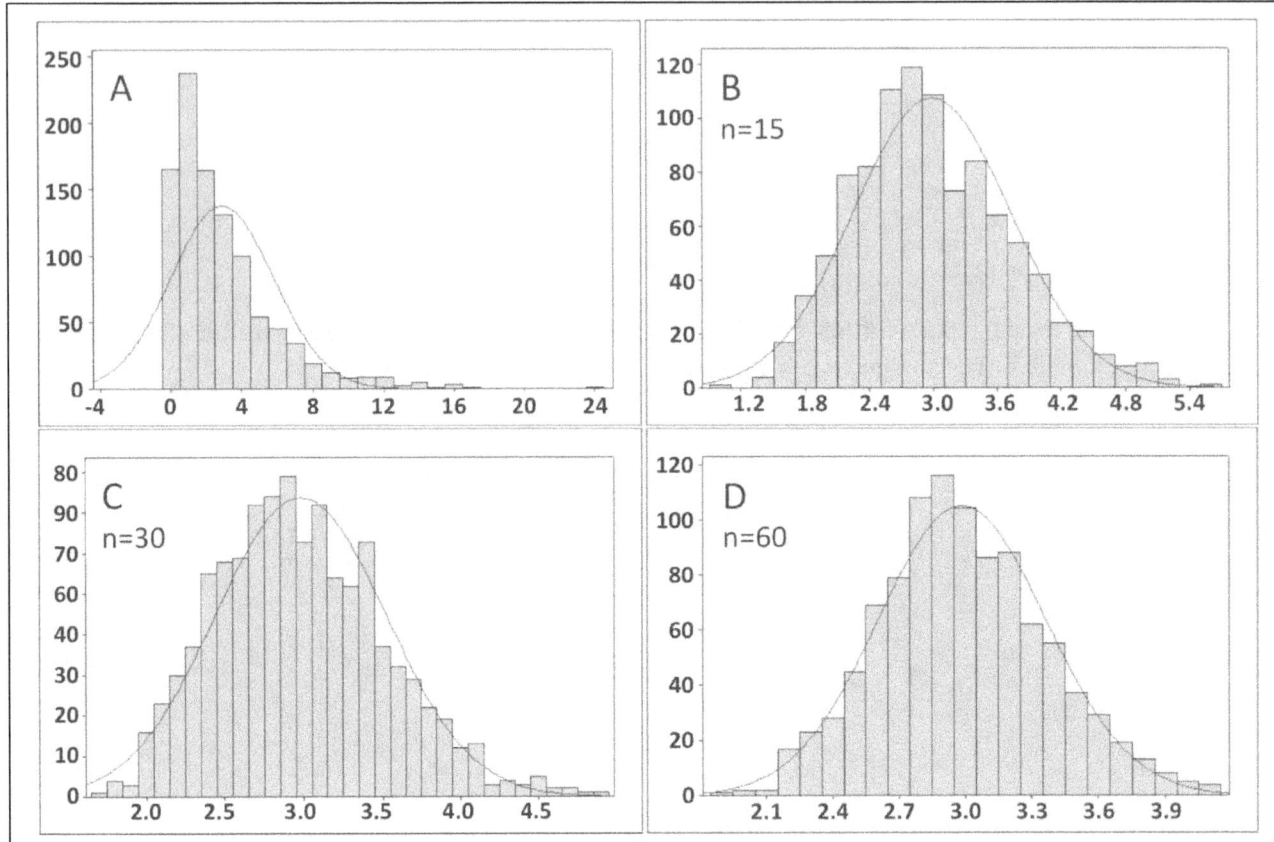

x_j are drawn from exponential distribution. As we increase n, we can see that the histogram of the mean values approaches normal distribution (from A to D).

CLT states that, given certain conditions, the arithmetic mean of a sufficiently large number of iterates of independent random variables, each with a well-defined expected value and well-defined variance, will be approximately normally distributed, regardless of the underlying distribution:

$$\lim_{n\to\infty} \frac{\bar{x} - \mu}{\sigma/\sqrt{n}} = N(0,1)$$

Why is this important? Many statistics have distributions that are approximately normal for large sample sizes, even when we are sampling from a distribution that is not normal. We can often use well-developed statistical inference procedures that are based on a normal distribution, even if we are sampling from a population that is not normal, provided we have a large sample size.

---

[5] The importance of CLT for QMC will be explained in the QMC section.

6. Monte Carlo Integration

6.1 Intro

MC methods were developed as a method for estimating integrals that could not be evaluated analytically. The first method for integration that comes to mind is to use a uniform grid and just count the squares below the curve. This method could be improved in various ways and there are other methods as well, but they all suffer a computational cost as the evaluating in d-dimensions and N grid-lines increases as N^d. An alternate and more efficient approach is to select points randomly, from a given probability distribution - the MC method. Thus, a straightforward application of MC is the evaluation of definite integrals.

First to define the arithmetic mean over N sample points:

$$\langle f \rangle = \frac{1}{N} \sum_{i=0}^{N-1} f(x_i) , \langle f^2 \rangle = \frac{1}{N} \sum_{i=0}^{N-1} f^2(x_i)$$

6.1.1 1D

For start, let us consider the 1D integral, which, by the mean value theorem, we can approximate by the interval multiplied by the mean of the function over that interval (the points x_i are selected randomly in the interval):

$$\int_a^b f(x)dx \approx (b-a)\langle f \rangle = (b-a)\frac{1}{N} \sum_{i=0}^{N-1} f(x_i) + O\left(\frac{1}{\sqrt{N}}\right)$$

By the CLT, the set of all possible sums over different $\{x_i\}$ will have a Normal distribution.

6.1.2 N-dimensions

- N - # of random points uniformly distributed in multidimensional volume - V.
- $x_0, \dots, x_{N-1}$ – random points in V.

The basic theorem of MC integration estimates the integral of function f over multidimensional volume:

$$\int f dV \approx V\langle f \rangle \pm \underbrace{V \sqrt{\frac{\langle f^2 \rangle - \langle f \rangle^2}{N}}}_{\text{Error term} \equiv \sigma_N}$$

$\pm$ means that the term is standard deviation error estimate for the integral. There is no guarantee that the error is distributed as a Gaussian, so the error term should be taken only as a rough indication of probable error.

6.1.3 2D Area

Suppose that we want to integrate g over region W that is not easy to sample randomly – W has a complicated shape. We find a region V that includes W and that can be easily sampled. Then we define f to be equal to g for points in W and equal to zero outside of W:

$$f(x) = \begin{cases} g(x), x \in W \\ 0, x \notin W \end{cases}$$

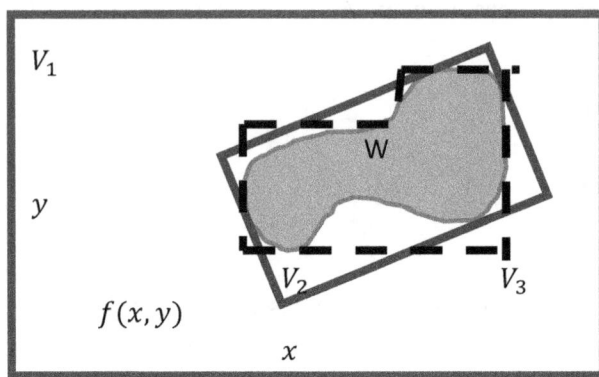

We want to try to make V enclose W as closely as possible, because the zero values of f will increase the error term – points outside W have no information content → the effective value of N is reduced.

V_1 is a poor choice; good choice V_2 can be sampled by picking a pair of uniform deviates (s, t) and map them onto (x, y) by a linear transformation. V_3 (dashed line) can be sampled by, using a uniform deviate, to choose between the left and right rectangular subregions and, then, using two more deviates to pick a point inside the chosen rectangle.

Let's create an object that embodies the described scheme – create `MCintegrate` object by providing (as constructor arguments) the following items:

- `xlo` – vector of lower limits of the coordinates for the box to be sampled,
- `xhi` – upper limits,
- `funcs` – vector-valued function that returns as its components one or more functions that we want to integrate simultaneously,
- Boolean function that returns whether a point is in the region W; the point will already be within V defined by `xlo` and `xhi`,
- Mapping function or `NULL` if there is no mapping function,
- Seed for the random number generator.

It is in `mcintegrate.h` under `struct MCintegrate`.

6.1.4 3D Volume - Example - Torus

We want to find the weight and position of the center of mass of a torus that intersects with faces of the box that encloses it. The object is defined by:

- $z^2 + \left(\sqrt{x^2 + y^2} - 3\right)^2 \leq 1$ (torus centered on the origin with radius = 3, minor radius = 1),
- $x \geq 1, y \geq -3$ (two faces of the box),
- $\rho = 1$ (constant density).

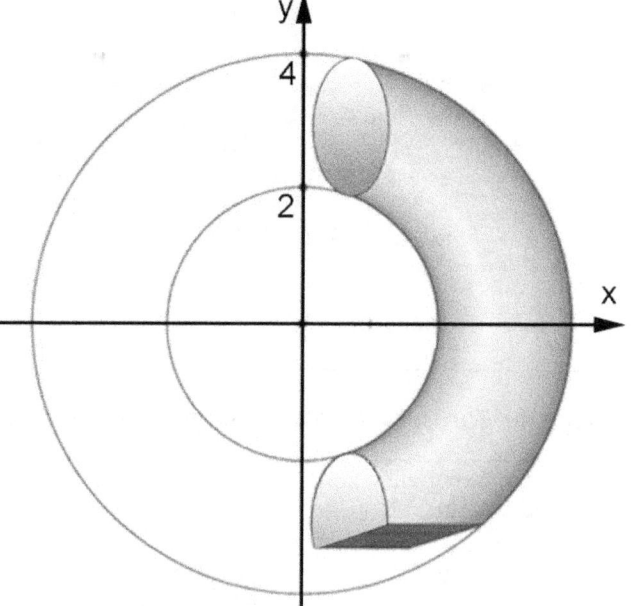

We want to estimate:

- Volume: $\int \rho \, dx \, dy \, dz$
- Coordinates of the center of the mass:
  $\frac{\int \{x,y,z\} \rho \, dx \, dy \, dz}{\int \rho \, dx \, dy \, dz}$; $\int \{x, y, z\} \rho \, dx \, dy \, dz$ – linear moments.

Code

In `MCintegrate.h` `calcanswer()`, `step()` and `nstep` are used to calculate:

$$\int f \, dV \approx V\langle f \rangle \pm V \sqrt{\frac{\langle f^2 \rangle - \langle f \rangle^2}{N}}, \text{ and } \langle f \rangle = \frac{1}{N} \sum_{i=0}^{N-1} f(x_i), \langle f^2 \rangle = \frac{1}{N} \sum_{i=0}^{N-1} f^2(x_i)$$

In `MC1.cpp` in main function we first create two double vectors with:

VecDoub xlo(3), xhi(3);[6]

MCintegrate.h

```
struct MCintegrate { //Object for Monte Carlo integration of one or more functions in an
ndim-dimensional region
        Int ndim,nfun,n; //number of dimensions, functions, points sampled
        VecDoub ff,fferr; //Answers: The integrals and their standard errors
        VecDoub xlo,xhi,x,xx,fn,sf,sferr;
        Doub vol; //Volume of the box V

        VecDoub (*funcsp)(const VecDoub &); //Pointers to the user-supplied functions
        VecDoub (*xmapp)(const VecDoub &);
        Bool (*inregionp)(const VecDoub &);
        Ranq1 ran; //Random number generator
```

---

[6] What is `VecDoub`? In nr3.h it is defined as typedef const NRvector<Doub>. There is also one for int.

So What is NRvector? #define NRvector vector

```
        //Constructor. The arguments are in the order described in the itemized list
below.
        MCintegrate(const VecDoub &xlow, const VecDoub &xhigh, //vectors xlo and xhi -
lower and upper limits of the coordinates for the rectangular box to be sampled. They
define V.
        VecDoub funcs(const VecDoub &), //funcs is a vector valued function that returns
as its components one or more functions that we want to integrate simultaneously
        Bool inregion(const VecDoub &), //return whether a point is in the region W that
we want to integrate. The point will already be in V.
        VecDoub xmap(const VecDoub &), Int ranseed);

        void step(Int nstep); //Sample an additional nstep points, accumulating the
various sums

        void calcanswers(); //Calculate answers ff and fferr using the current sums
};
```

Let's see the Torus. First write functions that describe the integrands:

```
VecDoub torusfuncs(const VecDoub &x) {//Return the integrands for Torus with rho=1
        Doub den = 1.; //den = density
        VecDoub f(4); //f will be with 4 dimensions: 0,1,2,3
        f[0] = den; //zeroth member
        for (Int i = 1; i<4; i++) f[i] = x[i - 1] * den; //x_i is the coordinates
x[0]=f[1]=x,x[0]=f[2]=y,x[0]=f[3]=z; multiply each position with weight-density
        return f;
}

Bool torusregion(const VecDoub &x) {
        return SQR(x[2])+SQR(sqrt(SQR(x[0])+SQR(x[1]))-3.) <= 1.; //Function of torus in
3D
}
```

MC1.cpp

The region of integration W inside V:

xlo and **xhi** are diagonal points of a 3D box, together they form a vector across that diagonal.

```
#include "stdafx.h"
#include <iostream>
#include "nr3.h"
#include "ran.h"
#include "mcintegrate.h"
using namespace std;


int main()
{
        // Unmapped code
        VecDoub xlo(3), xhi(3); //give xlo and xhi 3D
        xlo[0] = 0.; xhi[0] = 4.; // x coordinates
        xlo[1] = 0.; xhi[1] = 4.; // y coordinates
        xlo[2] = 0.; xhi[2] = 4.; // z coordinates
        MCintegrate mymc(xlo, xhi, torusfuncs, torusregion, NULL, 10201);
        mymc.step(1000000);
        mymc.calcanswers();
        // Display results
```

```
        cout << "\nmymc.ff = " << mymc.ff; // Display integral
        cout << "\nmymc.fferr = " << mymc.fferr; //Display error
    return 0;
}
```

Change of Variables

Example: $\rho = e^{5z}$. One way to include density is to change in `mcintegrate.h` from:

Doub den = 1.;

to

Doub den = exp(5.*x[2]);

This is not the best choice, because $\rho(z < -1) \approx 0 \rightarrow$ sampled points with $z < -1$ contribute almost nothing. A change of variable solves this problem:

$$ds = e^{5z}dz \Rightarrow s = \frac{1}{5}e^{5z} \Rightarrow z = \frac{1}{5}\ln 5s \Rightarrow \rho dz = ds$$

The limit for z for the V we gave is: $z \in (-1,1)$ becomes $.00135 < s < 29.682$.

We can correct this by setting argument for `xmap` from `NULL` to `torusmap`.

`torusmap` contents in `MCintegrate.h`:

```
VecDoub torusmap(const VecDoub &s) { //Return the mapping from s to z, mapping the other
coordinates by the identity map
    VecDoub xx(s);
    xx[2] = 0.2*log(5.*s[2]);
    return xx;
}
```

When the argument `xmap` is not `NULL`, it will assume that the sampling region defined by `xlo` an `xhi` is not in physical space, but rather needs to be mapped into physical space before either the functions or the region boundary are calculated $\rightarrow$ instead of changing `torusfuncs` or `torusregion` we can modify `xlo` and `xhi` and supply the function for `xmap`. The code in `MC1.cpp` is now:

```
    // Mapped code
    VecDoub slo(3), shi(3); //give slo (mapped xlo) and shi (mapped xhi)
    slo[0] = 1.; shi[0] = 4.; // mapped x coordinates
    slo[1] = -3.; shi[1] = 4.; // mapped y coordinates
    slo[2] = -1.; shi[2] = 1.; // mapped z coordinates
    MCintegrate mymc(slo, shi, torusfuncs, torusregion, torusmap, 10201); //now xmap
is not NULL
```

This method worked because the part of the integrand that we wanted to eliminate (e^{5z}) was both integrable and had an integral that could be inverted. In general this will not hold, and we must pull out of the integrand the best factor for which it holds. The criterion is to try to reduce the remaining integrand to a function that is as close as possible to constant. This and making W as close to V as possible is called *reduction of variance*.

The main disadvantage of Simple MC integration is that its accuracy increases only as $\sqrt{N}$.

6.2 Quasi- (Sub-) Random Sequences

Sequences of n-tuples that fill n-space more uniformly than uncorrelated random points are quasi-random sequences. The sample points are maximally avoiding of each other.

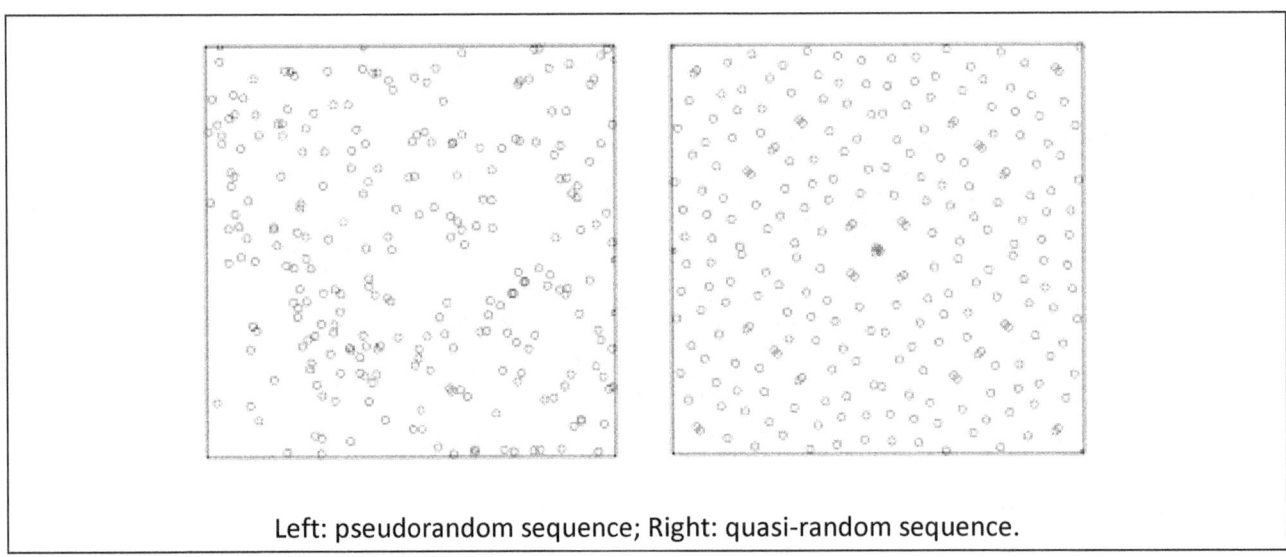

Left: pseudorandom sequence; Right: quasi-random sequence.

The advantage is the speed: error decreases as $1/N$, while with the Simple MC method it decreases as $1/\sqrt{N}$. However, quasi-random sequences have several drawbacks which make Quasi-MC hardly usable for QMC.[7]

6.3 Adaptive and Recursive MC Methods

These techniques are a *reduction of variance* - a procedure used to increase the precision of the estimates that can be obtained for a given number of iterations. Every output random variable from the simulation is associated with a variance which limits the precision of the simulation results. In order to obtain a greater precision and smaller confidence intervals for the output random variable of interest, reduction of variance techniques can be used.

6.3.1 Importance Sampling

Simple MC integration can suffer from low efficiency. Many functions of interest have significant weight in only a few regions. For example, most of the contributions to an integral of a Normal distribution are located near the central peak. In a simple MC, points are sampled uniformly, wasting considerable effort sampling the tails of the Normal distribution. The technique for overcoming this problem acts to increase the density of points in regions of interest and is called the importance sampling.

This is a general technique for estimating properties of a particular distribution, while only having samples generated from a different distribution than the distribution of interest. The basic idea is to change the probability measure so that the estimation is easier.

---

[7] There is a solution to a problem called *Randomized Quasi-MC method*.

General

Let's say that we have PDF p and f and we want to find out $\langle f \rangle$. x_i are from distribution p. However, as it can be seen from the graph, if we sample points $x_i \sim p$, most of the points will fall in the region where f is close to zero, and the points give us no information on the left region, the one that matters. Even if p had been uniformly distributed, the points to the right would be wasted. As a rule of a thumb, pick q similar to p.

```
Plot[{Evaluate@Table[PDF[NormalDistribution[μ, σ], x], {μ, {3, 7}}, {σ, {1, 1}}],
   Evaluate@Table[.2 * CDF[NormalDistribution[μ, 1], x - 4] - .2, {μ, 1}]},
 {x, 0, 10}, Filling → Axis]
```

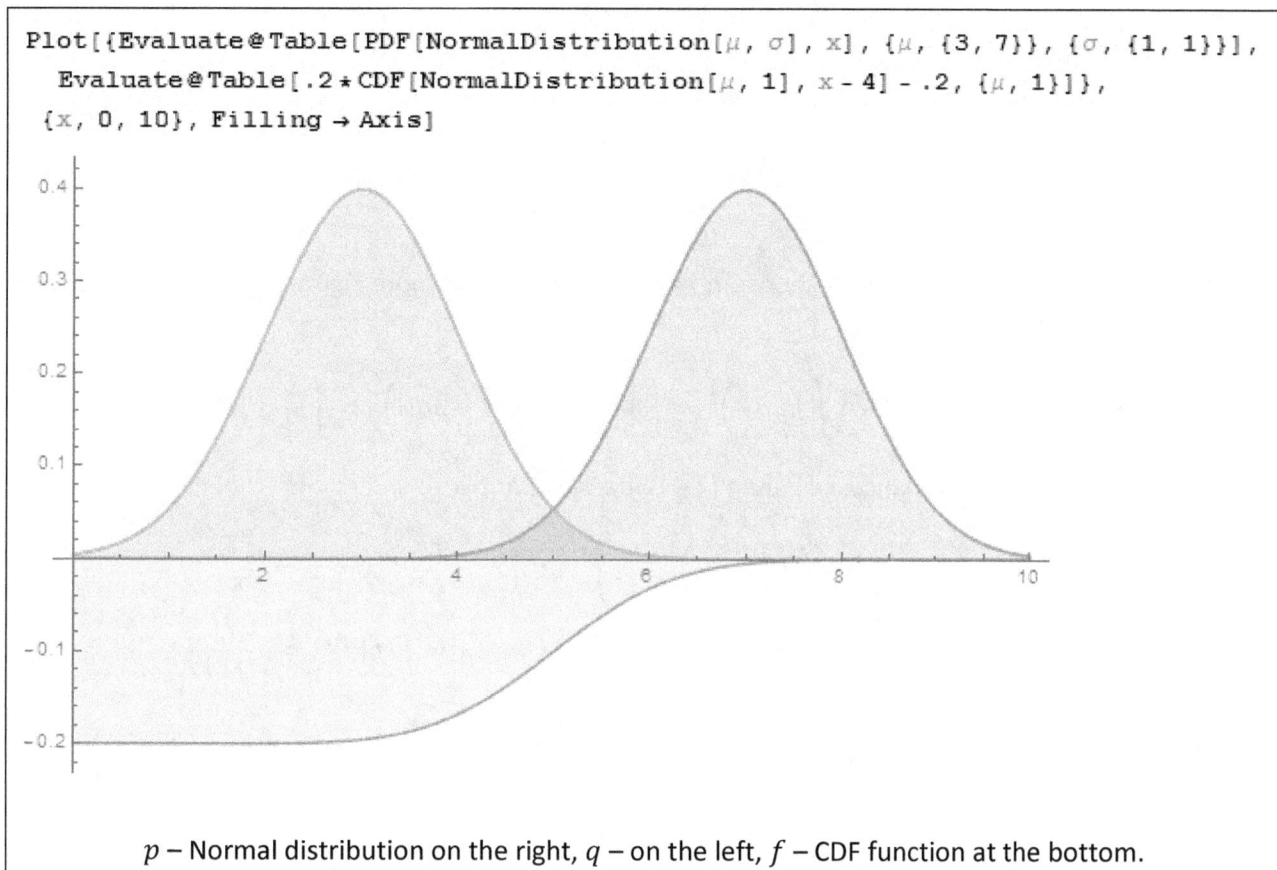

p – Normal distribution on the right, q – on the left, f – CDF function at the bottom.

In Simple MC integration: $\langle f \rangle = \frac{1}{N}\sum_{i=0}^{N-1} f(x_i)$, $x_i \sim p$, and $\sigma^2(MC) = \frac{1}{N}\sigma^2(f(x))$

With importance sampling:

$$\langle f \rangle = \int f(x)p(x)dx = \int \left(f(x)\frac{p(x)}{q(x)}\right)q(x)dx \underset{x_i \sim q}{\approx} \frac{1}{N}\sum_{i=0}^{N-1} f(x_i)\frac{p(x_i)}{q(x_i)}$$

$w(x) = \frac{p(x)}{q(x)}$ – importance weight function; $w(x_i) = \frac{p(x_i)}{q(x_i)}$ – importance weight.

$$\langle f \rangle \approx \frac{1}{N}\sum_{i=0}^{N-1} f(x_i)\,w(x_i)$$

Reduced variance by Importance Sampling: $\sigma^2(IS) = \frac{1}{N}\sigma^2(f(x)w(x))$

This is how we lowered the expected error of our estimate.

Multidimensional Volume

$$\int f \, dV = \int \frac{f}{p} p \, dV = \left| h = \frac{f}{p} \right| = \int h p \, dV$$

We can integrate f by sampling h; however, not with uniform probability density dV but rather with nonuniform density $p \, dV$. Suppose that the points $x_i \in V$ with a probability density p satisfying: $\int p \, dV = 1$. The integral of f is estimated using N sample points $\{x_i\}, i = 0, \dots, N-1$:

$$\int f \, dV = \int h p \, dV \approx \langle h \rangle \pm \underbrace{\sqrt{\frac{\langle h^2 \rangle - \langle h \rangle^2}{N}}}_{\substack{\text{st.deviation} \\ \text{error estimate}}}$$

The best choice for sampling p is to choose $h \approx const$. We can be more rigorous by focusing on the variance per sample point:

$$S = \langle h^2 \rangle - \langle h \rangle^2 \approx \int h^2 p \, dV - \left(\int h p \, dV \right)^2 = \int \frac{f^2}{p^2} p \, dV - \left(\int \frac{f}{p} p \, dV \right)^2 = \int \frac{f^2}{p} \, dV - \left(\int f \, dV \right)^2$$

We want to find minimum variance, i.e. when (λ – Lagrange multiplier):

$$\frac{\delta S}{\delta p} = 0 = \frac{\delta}{\delta p} \left(\int \frac{f^2}{p} \, dV - \left(\int f \, dV \right)^2 + \lambda \int p \, dV \right) = -\frac{f^2}{p^2} + \lambda \Rightarrow p = \frac{|f|}{\sqrt{\lambda}}$$

$$= \left| \int p \, dV = 1 \Rightarrow \int \frac{|f|}{\sqrt{\lambda}} \, dV = 1 \Rightarrow \sqrt{\lambda} = \int |f| \, dV \right| = \frac{|f|}{\int |f| \, dV}$$

$$S_{\text{optimal}} = \int \frac{f^2}{p} \, dV - \left(\int f \, dV \right)^2 = \int \frac{f^2}{\frac{|f|}{\int |f| \, dV}} \, dV - \left(\int f \, dV \right)^2 = \int |f| \, dV \int |f| \, dV - \left(\int f \, dV \right)^2$$

$$= \left(\int |f| \, dV \right)^2 - \left(\int f \, dV \right)^2$$

$p \propto |f|$ is optimal.

Case: If f takes on a known constant value in most of the V, it's a good idea to add a constant so as to make that value zero. Then the accuracy by importance sampling depends not on how small S_{optimal} is, but rather on how small is S for an implementable p.

6.3.2 Stratified Sampling
Consider the V divided into sub-regions – stratification – in order to try to get a smaller variance. The optimal allocation of sample points among the regions is to have the number of points in each region j proportional to the square root of the variance of the function in that sub-region: $j \propto \sigma_a$. In practice this is not useful for spaces with high dimensionality.

Sampling two sub-volumes

- $\langle f \rangle = \frac{1}{N} \Sigma_i f(x_i)$ – Simple (Uniformly sampled) MC estimator of the average,
- $\langle\langle f \rangle\rangle = \frac{1}{V} \int f \, dV$ – true average.
- $\text{Var}(\langle f \rangle)$ - The variance of the estimator, measures the square of the error of the MC integration.
- $\text{Var}(f) = \langle\langle f^2 \rangle\rangle - \langle\langle f \rangle\rangle^2$ – The variance of f.

$$\text{Var}(\langle f \rangle) = \frac{\text{Var}(f)}{N}$$

We divide V into equal sub-volumes a and b, and sample $N/2$ points in each. The other estimator for a true average, other than the one described: $\langle f \rangle' = \frac{1}{2}(\langle f \rangle_a + \langle f \rangle_b)$

$\text{Var}_a(f) = \langle\langle f^2 \rangle\rangle_a - \langle\langle f \rangle\rangle_a^2, \text{Var}_b(f) = \langle\langle f^2 \rangle\rangle_b - \langle\langle f \rangle\rangle_b^2$ – Variance in sub-regions.

$$\text{Var}(\langle f \rangle') = \frac{1}{4}\left(\text{Var}(\langle f \rangle_a) + \text{Var}(\langle f \rangle_b)\right) = \frac{1}{4}\left(\frac{\text{Var}_a(f)}{N/2} + \frac{\text{Var}_b(f)}{N/2}\right) = \frac{1}{2N}\left(\text{Var}_a(f) + \text{Var}_b(f)\right)$$

Therefore, by the parallel axis theorem: $\text{Var}(f) = \frac{1}{2}\left(\text{Var}_a(f) + \text{Var}_b(f)\right) + \frac{1}{4}(\langle\langle f \rangle\rangle_a - \langle\langle f \rangle\rangle_b)^2$

Comparing $\text{Var}(\langle f \rangle)$, $\text{Var}(\langle f \rangle')$ and $\text{Var}(f)$, one sees that the stratified (into two sub-volumes) sampling gives a variance that is never larger than the Simple MC case – and smaller whenever the means of the stratified samples, $\langle\langle f \rangle\rangle_a$ and $\langle\langle f \rangle\rangle_b$, are different: $\text{Var}(\langle f \rangle') \leq \text{Var}(\langle f \rangle)$

Sampling two sub-volumes with different number of points

$N = N_a + N_b \rightarrow \text{Var}(\langle f \rangle') = \frac{1}{4}\left(\frac{\text{Var}_a(f)}{N_a} + \frac{\text{Var}_b(f)}{N_b}\right)$

Notation: $\sigma_a = \sqrt{\text{Var}_a(f)}$

The variance is minimal when: $\frac{N_a}{N} = \frac{\sigma_a}{\sigma_a + \sigma_b} \rightarrow \text{Var}(\langle f \rangle') = \frac{(\sigma_a + \sigma_b)^2}{4N}$.

6.3.3 Mixed Strategies

Importance sampling:

- Concentrates sample points where the magnitude of the integrand $|f|$ is largest.
- Depends on already knowing some approximation to the integral, so that we are able to generate random points x_i with the desired probability density p.
- To the extent that p is not ideal, we are left with an error that decreases as $\frac{1}{\sqrt{N}}$.
- It is bad in the region where f is changing rapidly, in which case it works by smoothing the values of h and is effective only to the extent that we succeed in this.

Stratified sampling:

- Concentrates sample points where the $\sigma^2(f)$ is largest.
- Works by smoothing out the fluctuations of the number of points in sub-regions, not by smoothing the values of points.
- Dividing V into N equal sub-regions – the simples stratification – gives a method whose error decreases as N^{-1}.

- If the integrand is negligible in all but a simple sub-region, then the resulting one-sample integration is useless.
- It is useful when we can estimate variances, so that we can put unequal numbers of points in different sub-regions, according to $\frac{N_a}{N} = \frac{\sigma_a}{\sigma_a + \sigma_b}$, and if we can find a way of dividing a region into practical number of sub-regions.

These two are not incompatible. In most cases of interest f is small everywhere in V but in small fractional volume of active regions. In these regions $|f|$ and standard deviation σ are comparable in size, so both techniques will give about the same concentration of points. It is also possible to nest the two techniques.

6.3.4 Adaptive MC: VEGAS

It is widely used for multidimensional integrals that occur in elementary particle physics.

The VEGAS algorithm is a method for error reducing in MC simulations by using a known or approximate probability distribution function to concentrate the search in those areas of the integrand that make the greatest contribution to the final integral. It is based on importance sampling, it samples points from the probability function so that the points are concentrated in the regions that make the greatest contribution. It approximates the exact distribution by making a number of passes over the integration region while histogramming the function. Each histogram is used to define a sampling distribution for the next pass. Asymptotically this procedure converges to the desired distribution. In order to avoid the number of histogram bins growing like K^d (K – number of sub-divisions of the independent variable, d – number of dimensions of the probability distribution), the probability distribution is approximated by a separable function, so that the number of bins required is only Kd. This is equivalent to locating the peaks of the function from the projections of the integrand onto the coordinate axes and the efficiency of VEGAS depends on the validity of this assumption. It is most efficient when the peaks are well-localized. If an integrand can be rewritten in a form which is approximately separable this will increase the efficiency of integration.

Method

The basic technique for importance sampling in VEGAS is to construct, adaptively, a multidimensional weight function g that is separable:

$$p \propto g(x, y, z \ldots) = g_x(x) g_y(y) g_z(z) \ldots$$

Such a function avoids the K^d explosion in two ways:

i. It can be stored in the computer as d separate 1D functions, each defined by K tabulated values - Kd replaces K^d.
ii. It can be sampled as a probability density by consecutively sampling the d 1D functions to obtain coordinate vector components $(x, y, z, \ldots)$.

Given a set of g functions (initially all constant), one samples f, accumulating the overall estimator of the integral and Kd estimators. These then determine improved g functions for the next iteration.

The optimal separable weight function:

$$g_x \propto \sqrt{\int \frac{f^2(x, y, z, \dots)}{g_y g_z \dots} \, dydz \dots}$$

In 1D this reduces to $g \propto |f|$.

When f is concentrated in one or a few regions, then g's quickly become large at coordinate values that are the projections of these regions onto the coordinate axes. The accuracy of MC is then enormously enhanced over simple MC.

Weakness of VEGAS:

- If the projection of f onto coordinates is uniform, it gives no concentration of sample points in those dimensions.
- The worst case is an integrand concentrated close to a body diagonal line $(0,0,0 \dots)$ to $(1,1,1 \dots)$, since this geometry is completely non-separable, and we get no advantage.
- It will not do well when the integrand is concentrated in curved trajectories or hypersurfaces unless they happen to be oriented close to the coordinate directions.

While statistically independent, iterations assist each other, since each one is used to refine the sampling grid for the next one. The results of all iterations are combined into a single best answer, and its estimated error (m – degrees of freedom):

$$I_{\text{best}} = \frac{\sum_{i=0}^{m-1} I_i / \sigma_i^2}{\sum_{i=0}^{m-1} 1/\sigma_i^2}, \sigma_{\text{best}} = \frac{1}{\sqrt{\sum_{i=0}^{m-1} \frac{1}{\sigma_i^2}}}$$

$$\frac{\chi^2}{m} = \frac{1}{m-1} \sum_{i=0}^{m-1} \frac{(I_i - I_{\text{best}})^2}{\sigma_i^2}$$

$\frac{\chi^2}{m} \gg 1 \rightarrow$ the results of the iterations are statistically inconsistent, and the answers are suspect.

In most cases there is no need for more than 5 or 10 iterations.

Code

- K = NDMX – the maximum number of increments along each axis,
- $\max d$ = MXDIM,
- m = itmx – number of statistically independent evaluations of the integral - iterations, each with
- N = ncall – function evaluations,
- d = ndim,
- fxb – user supplied function over a rectangular volume specified by
- regn[0..2*ndim-1] – a vector consisting of ndim lower-left and upper-right coordinates,
- init – input flag that signals whether this call is a new start or a subsequent call for additional iterations,
- nprn – input flag (normally 0), controls the amount of diagnostic output,
- tgral – returned answers, the best estimate of the integral,

- `sd` – the best standard deviation,
- `chi2a` – an indicator of whether consistent results are being obtained,
- w_i = `wgt`.

The input flag `init` can be used to advantage. One might have a call with `init=0, ncall=1000, itmx=5` immediately followed by a call with `init=1, ncall=10000, itmx=1`. The effect would be to develop a sampling grid over 5 iterations of a samples, then do a single high accuracy integration on the optimized grid.

The `fxn` has an argument `wgt` in addition to the expected evaluation point x. In most applications we can ignore `wgt` inside the function. Occasionally, however, we might want to integrate some additional function along with the f. The integral of any such function g can be estimated by:

$$I_g = \sum_i w_i g(x)$$

vegas.h

```
void vegas(VecDoub_I &regn, Doub fxn(VecDoub_I &, const Doub), const Int init,
        const Int ncall, const Int itmx, const Int nprn, Doub &tgral, Doub &sd,
        Doub &chi2a) {
        /* Performs MC integration of a user-supplied ndim-dimensional function fxn over
        a rectangular volume specified by regn[0..2*ndim-1], a vector consisting of ndim
        lower-left coordinates of the region followed by ndim upper-right coordinates.
        The integration consists of itmx iterations, each with approximately ncall calls
        to the function. After each iteration the grid is refined; more than 5 or 10
        iterations are rarely useful. The input flag init signals whether this call is
        a new start or a subsequent call for additional iterations (see comments in the
        code). The input flag nprn (normally 0) controls the amount of diagnostic
        output. Returned answers are tgral (the best estimate of the integral),
        sd (its standard deviation), and chi2a (chi^2 per degree of freedom, an
        indicator of whether consistent results are being obtained). */
                                            .
                                            .
                                            .
        static Ranq1 ran_vegas(RANSEED); /*Initialize a captive, static random
                                        generator; default is RAN, but Ranq1 is faster*/

        Int ndim=regn.size()/2;
        if (init <= 0) { //Normal entry. Enter here on a cold start
                mds=ndo=1; //Change to mds=0 to disable stratified sampling, i.e. use
importance sampling only
                for (j=0;j<ndim;j++) xi[j][0]=1.0;
        }
        if (init <= 1) si=swgt=schi=0.0;
        //Enter here to inherit the grid from a previous call, but not its answers
        if (init <= 2) { //Enter here to inherit the previous grid and its answers
                nd=NDMX;
                ng=1;
                if (mds != 0) { //Set up for stratification
                                            .
                                            .
                                            .
        for (it=0;it<itmx;it++) {
                /*Main iteration loop. Can enter here (init>=3) to do an additional itmx
```

```
            iterations with all other parameters unchanged */
                                    .
                                    .
                                    .
            if (mds < 0) { //Use stratified sampling
                                    .
                                    .
                                    .
        tsi *= dv2g; //Compute final results for this iteration
                                    .
                                    .
                                    .
        for (j=0;j<ndim;j++) { //Refine the grid
                                    .
                                    .
                                    .
```

VEGAS.cpp

```cpp
// Computes the Volume of the Torus in the Box using VEGAS

#include "stdafx.h"
#include <iostream>
#include "nr3.h"
#include "ran.h" //Must be included before rebin.h and vegas.h
/*rebin.h is Utility routine used by vegas to rebin a vector of
densities contained in row j of xi into new bins defined by a vector r*/
#include "rebin.h" //It is important to include it before vegas.h
#include "vegas.h"

using namespace std;
Doub torusfunc(const VecDoub &x, const Doub wgt) {
        Doub den = 1.; //exp(5.*x[2]);
        //Torus equation
        if (SQR(x[2]) + SQR(sqrt(SQR(x[0]) + SQR(x[1])) - 3.) <= 1.) return den;
        else return 0.;
}

int main()
{
        Doub tgral, sd, chi2a;
        VecDoub regn(6); //lower-left and upper-left coordinates of the Box
        /*regn[0] = 1.; regn[3] = 4.; //x values
        regn[1] = -3.; regn[4] = 4.; //y values
        regn[2] = -1.; regn[5] = 1.; //z values*/
        regn[0] = 0.; regn[3] = 4.; //x values
        regn[1] = 0.; regn[4] = 4.; //y values
        regn[2] = 0.; regn[5] = 4.; //z values
        //vegas(regn, fxn, init, ncall, itmx, nprn, tgral, sd, chi2a)
        vegas(regn, torusfunc, 0, 20000, 10, 0, tgral, sd, chi2a);
        vegas(regn, torusfunc, 1, 1000000, 2, 0, tgral, sd, chi2a);
}
```

Output
```
Input parameters for vegas  ndim=    3  ncall=    9826
                            it=    0  itmx=    10
```

46

```
                                nprn=      0  ALPH=        1.5
                                 mds=      1  nd=     50
                            x1[ 0]=              1 xu[ 0]=           4
                            x1[ 1]=             -3 xu[ 1]=           4
                            x1[ 2]=             -1 xu[ 2]=           1
iteration no.    1 : integral =        217.066 +/-    2.83259
all iterations:     integral =    217.066+-   2.83259 chi**2/IT n =         0
iteration no.    2 : integral =        219.692 +/-    1.2885
all iterations:     integral =    219.242+-   1.17286 chi**2/IT n = 0.711599
iteration no.    3 : integral =        220.338 +/-    1.22622
all iterations:     integral =    219.765+-  0.847571 chi**2/IT n = 0.564616
iteration no.    4 : integral =          218.3 +/-    1.21184
all iterations:     integral =    219.284+-   0.69455 chi**2/IT n = 0.703539
iteration no.    5 : integral =        217.752 +/-    1.24046
all iterations:     integral =    218.918+-  0.606021 chi**2/IT n = 0.818099
iteration no.    6 : integral =        217.745 +/-     1.2494
all iterations:     integral =    218.695+-  0.545263 chi**2/IT n = 0.797358
iteration no.    7 : integral =        219.626 +/-    1.22294
all iterations:     integral =    218.849+-  0.498005 chi**2/IT n = 0.745016
iteration no.    8 : integral =        219.442 +/-    1.25004
all iterations:     integral =     218.93+-  0.462642 chi**2/IT n = 0.666287
iteration no.    9 : integral =        218.127 +/-    1.25468
all iterations:     integral =    218.834+-  0.434073 chi**2/IT n = 0.628116
iteration no.   10 : integral =        219.386 +/-    1.28411
all iterations:     integral =    218.891+-  0.411215 chi**2/IT n = 0.576717
Input parameters for vegas  ndim=      3  ncall=    877952
                                  it=     10  itmx=      2
                                nprn=      0  ALPH=        1.5
                                 mds=     -1  nd=     38
                            x1[ 0]=              1 xu[ 0]=           4
                            x1[ 1]=             -3 xu[ 1]=           4
                            x1[ 2]=             -1 xu[ 2]=           1
iteration no.    1 : integral =        218.821 +/- 0.0733578
all iterations:     integral =    218.821+-0.0733578 chi**2/IT n =         0
iteration no.    2 : integral =        218.906 +/- 0.0597158
all iterations:     integral =    218.872+-0.0463114 chi**2/IT n = 0.802997
Input parameters for vegas  ndim=      3  ncall= 1.76947e+06
                                  it=      2  itmx=      1
                                nprn=      0  ALPH=        1.5
                                 mds=     -1  nd=     48
                            x1[ 0]=              1 xu[ 0]=           4
                            x1[ 1]=             -3 xu[ 1]=           4
                            x1[ 2]=             -1 xu[ 2]=           1
iteration no.    1 : integral =        218.902 +/- 0.0355378
all iterations:     integral =    218.891+-0.0281935 chi**2/IT n =   10574.6
```

Test VEGAS vs Simple MC

Let our torus be in a box $(0,0,0)$ and $(4,4,4)$, that is the $1/8$ of a complete torus is in that box. Set the density to 1. The Volume is exactly 7.4022.

- Simple MC results (`step= 1000000`): 7.40461 ± 0.0204711,
- VEGAS (`ncall={20000,1000000}`, `itmx={10,2}`): 7.40148 ± 0.00233.

Simple MC missed by 0.00241 and VEGAS by 0.00072, a clear victory for VEGAS.

7. Random movement and Markov Chain

7.1 Markov Chain

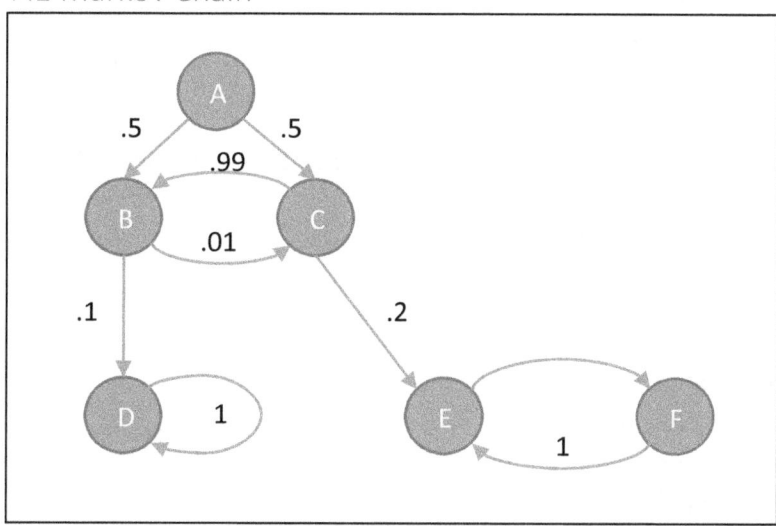

Example Transition Diagram

- Transient state – we can leave that state (state A, B and C). There exists a place where we can go and not come back. Transient state will be visited only for a finite number of times – we will exit them in the long run.
- Ephemeral – we can only leave (state A).
- Absorbing – we can never leave (state D).
- Recurrent – once we go in, we can only go between these states (states E and F). Starting from a state i there is a way of returning to i.

Transition probabilities go into the matrix P:

$$P = \begin{pmatrix} \begin{array}{c} \downarrow \text{We are} \\ \text{currentlly in} \end{array} \Big/ \begin{array}{c} \text{We are} \\ \text{going to} \rightarrow \end{array} & \begin{array}{cccccc} A & B & C & D & E & F \end{array} \\ \begin{array}{c} A \\ B \\ C \\ D \\ E \\ F \end{array} & \begin{array}{cccccc} 0 & .5 & .5 & 0 & 0 & 0 \\ 0 & 0 & .99 & .01 & 0 & 0 \\ 0 & .8 & 0 & 0 & .2 & 0 \\ 0 & 0 & 0 & 1 & 0 & 0 \\ 0 & 0 & 0 & 0 & 0 & 1 \\ 0 & 0 & 0 & 0 & 1 & 0 \end{array} \end{pmatrix}$$

Sum of probabilities for each row is 1, for every row: $\sum_j P_{ij} = 1$

$\pi^{(0)} = (1,0,0,0,0,0)$ – Initial state (we are at state A)

$$\pi^{(1)} = \pi^{(0)} P$$

$$\pi^{(n)} = \cdots = \pi^{(0)} P^n$$

E.g. calculating the probabilities:

$$P(X_1 = B, X_2 = C, X_3 = E | X_0 = A) = P_{AB} P_{BC} P_{CE}$$

$$P(X_4 = E | X_0 = A) = P_{AC} P_{CE} P_{EF} P_{FE} + P_{AC} P_{CB} P_{BC} P_{CE}$$

This is the sum of all the probabilities to go to the state E given initial state A.

The last equation could have been calculated using recursion equation, but for the small number of steps we can use brute force method.

Example. Customers are in queue. The store can accept at most 10 customers.

Cases:

- With each arrival, state moves for 1 higher,
- With each departure, state moves for 1 lower,
- There is a possibility that nothing happens,
- There is a possibility of having simultaneously arrival and departure.

Probabilities:

- p – probability for arrival at the state,
- q – probability for departure,
- $1 - p$ – no departure,
- $1 - q$ – no arrival,
- $q(1 - p)$ – departure, no arrival,
- $p(1 - q)$ – arrival, no departure.

At the ends:

- At 0, there can be no departure, only arrival or that it stays the same.
- At 10, there can be no arrival, only departure or that it stays the same.

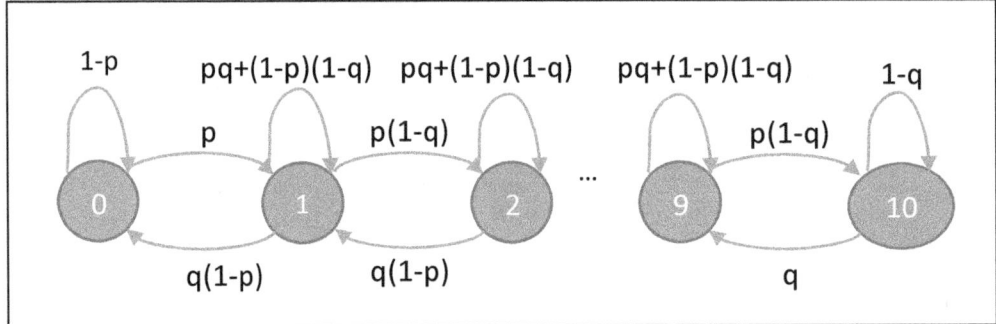

- P_{31} – probability of going from state 3 to state 1.
- P_{ij} – from state i to state j.

Markov property: the probability depends only on the current state. If we know everything that happened before the current state, that does not change anything.

$$P_{ij} = P(X_{n+1} = j | X_n = i) = P(X_{n+1} = j | X_n = i, X_{n-1}, \dots, X_0)$$

We need to choose what the state variable is, and define it in a way that includes all information that has been accumulated that has some relevance for the future.

Modes specification:

- Identify the possible states,
- Identify the possible transitions,
- Identify the transition probabilities.

7.1.1 n-step transition probabilities

Case 1

State occupancy probabilities given initial state i: $r_{ij}(n) = P(X_n = j | X_0 = i)$

With what probability will that state, after n transitions, be in state j?

$$r_{ij}(0) = \begin{cases} 1, i = j \\ 0, i \neq j \end{cases}, r_{ij}(1) = P_{ij}$$

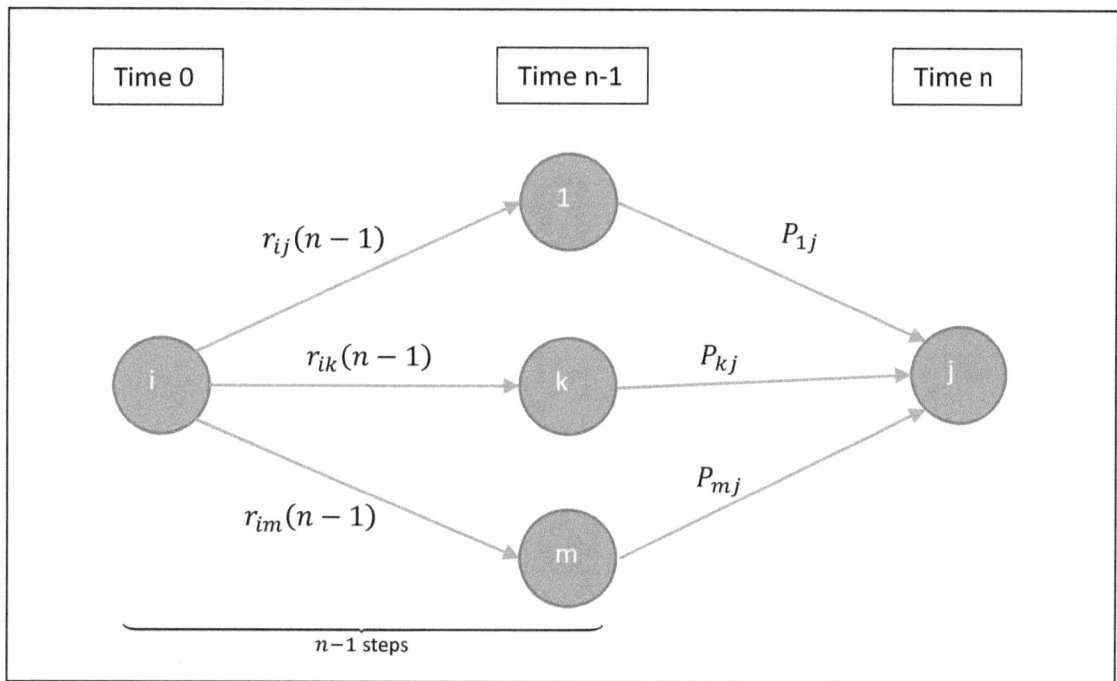

Markov property: No matter which path we took from $i \rightarrow 1$, P_{ij} does not depend on that path.

Key recursion: $r_{ij}(n) = \sum_{k=1}^{m} r_{ik}(n-1) P_{kj}$

Once we consider all scenarios, and we add the probabilities, we end up with key recursion for $(\forall i, j)$. It tells us that once we have computed the $(n-1)$ step transition probabilities, then we can compute the n step transition probability. This is a recursion that we execute for $(\forall i, j)$ simultaneously.

With random individual state:

$$P(X_n = j) = \sum_{i=1}^{m} P(X_0 = i) r_{ij}(n)$$

- $P(X_0 = i)$ – probability that the initial state is i,
- $r_{ij}(n)$ – probability that after n steps state is j.

Case 2

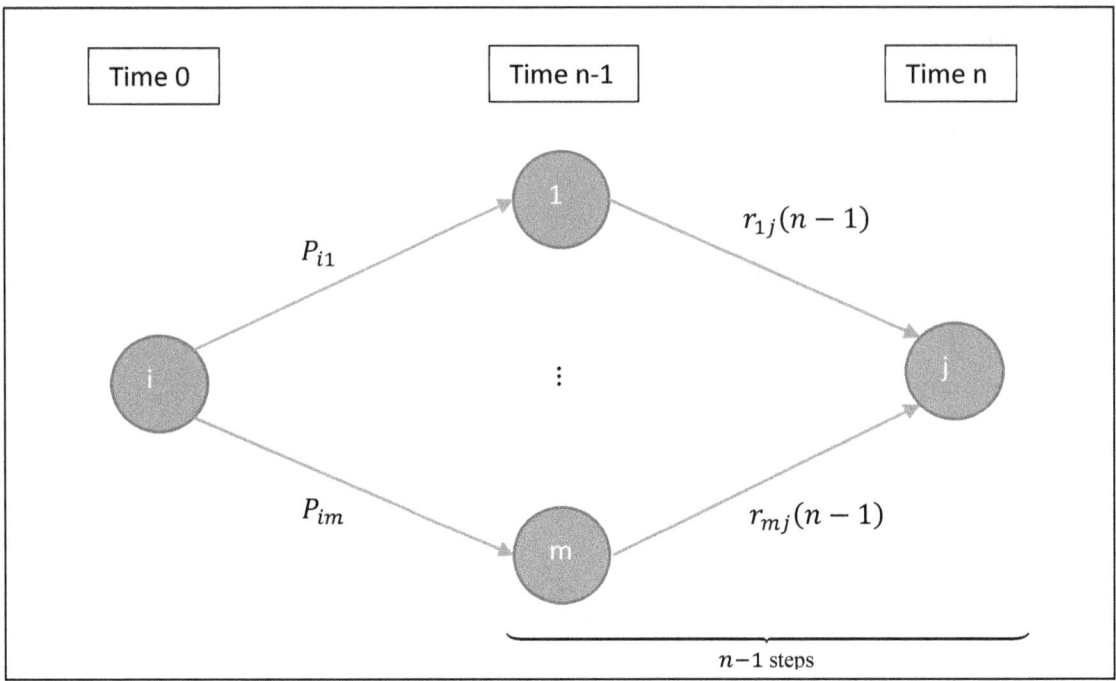

There are many paths from each state $\{1, \dots, m\}$ to j.

$$r_{ij}(n) = \sum_{k=1}^{m} P_{ik} r_{kj}(n-1)$$

We got similar equation to that from "Case 1".

Example

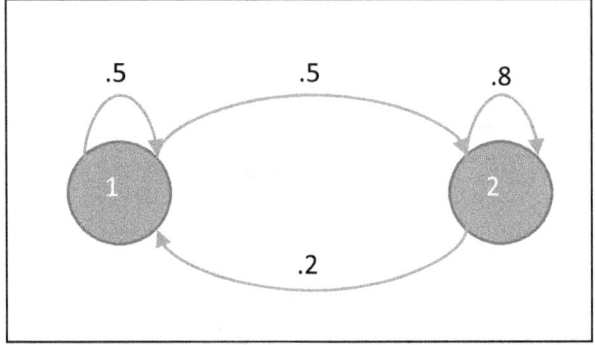

$$r_{11}(n) = \underbrace{r_{12}(n-1)}_{\substack{\text{At n-1 we}\\\text{can be at 2}}} \cdot \underbrace{.2}_{\substack{\text{from 2}\\\text{back to 1}}} + \underbrace{r_{11}(n-1)}_{\substack{\text{We stay}\\\text{where we are}}} \cdot .5$$

$$r_{12}(n) = 1 - r_{11}(n)$$

n	$r_{11}(n)$	$r_{12}(n)$	$r_{21}(n)$	$r_{22}(n)$
0	1	0	0	1
1	.5	.5	.2	.8
2	$.5^2 + .5 \cdot .2 = .35$	$1 - .35 = .65$		
100	$\approx 2/7$	$\approx 5/7$	$\approx 2/7$	$\approx 5/7$
101	$\approx 2/7$	$\approx 5/7$	$\approx 2/7$	$\approx 5/7$

Probabilities are the same no matter where we start at – the chain does not care about the initial chain in the long run. MCMC probabilities eventually enter a steady state.

7.1.2 Generic convergence questions

Is it always the case that for $n \to \infty$ the transition probabilities converge? If yes, is the limit not affected by the initial state? Usually both will be true, but there are cases when it is not.

Example: Convergence fails

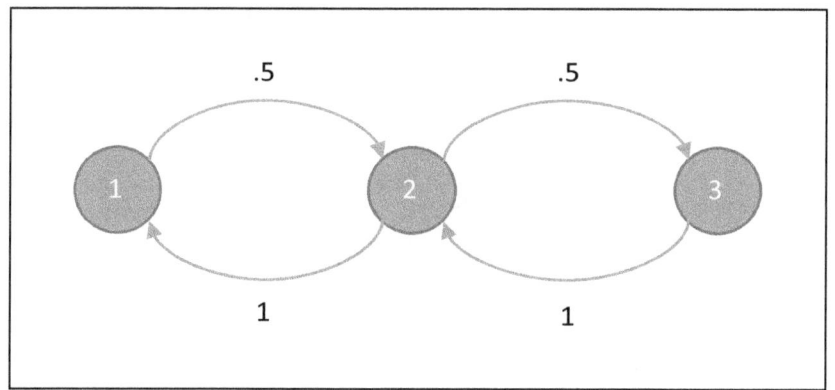

$$r_{22}(n) = \begin{cases} 1, & n \text{ even} \\ 0, & n \text{ odd} \end{cases}$$

Example: If we have convergence, does initial state matter?

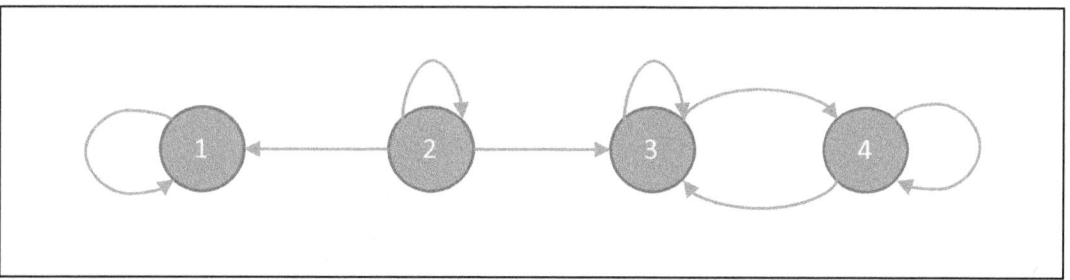

$$r_{11}(n) = 1$$

$$r_{31}(n) = 0$$

$$r_{21}(n) \to \frac{1}{2}, n \to \infty$$

Def: Recurrent class is a group of recurrent states that are connected.

For the initial state to not matter, we should not have multiple recurrent classes – we should only have one.

7.1.3 Periodic structure

Initial conditions matter if chain has periodic structure. The state space of a chain is periodic if we can group the states into a d number of clusters, and the transition diagram has a property that from a cluster we always make a transition into the next cluster.

Example 1

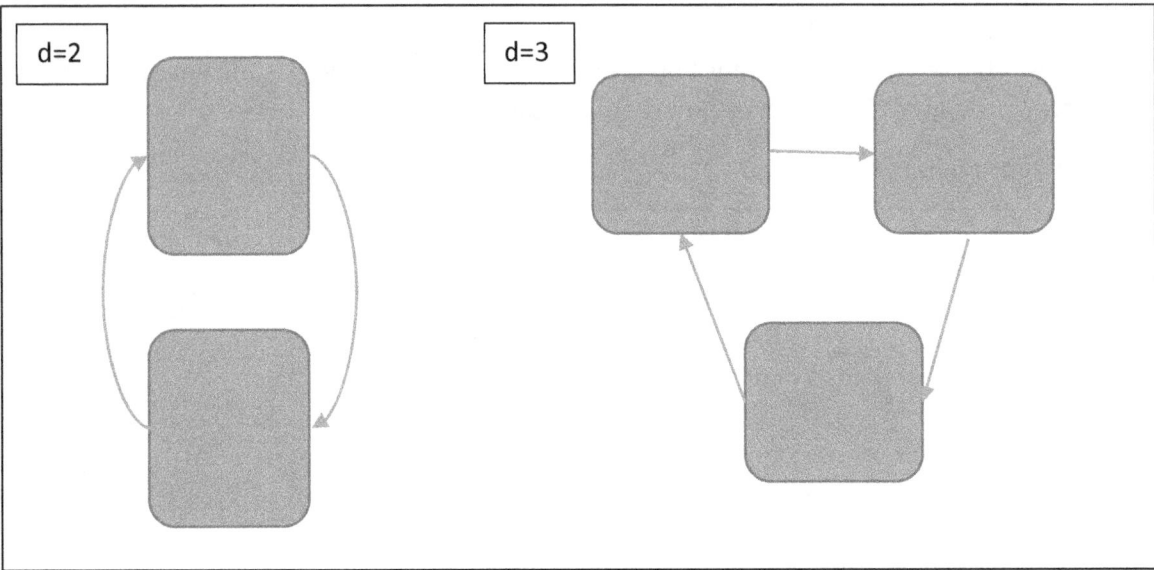

On the above example, for d=2, even or odd times matter → the exact time does matter in determining the probabilities of different states. The probability of being at the particular state cannot converge to a state value.

Example 2

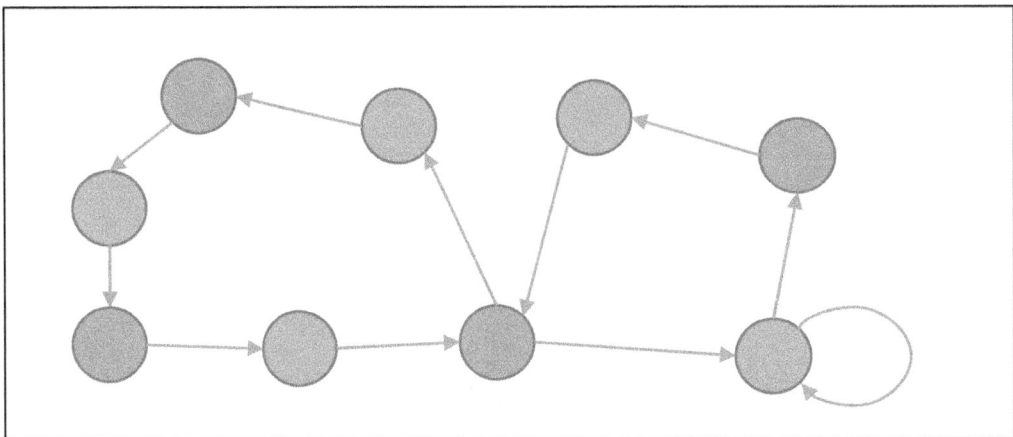

1. The states are not colored and there is no self-transition to the right. It might not be obvious that the chain is periodic.
2. Let us now apply color → we can see that there is periodicity of going to orange to blue and vice versa → we can group it into a cluster of orange states and cluster of blue states.
3. Let's add the self-transition → it is not periodic anymore since we are staying at the same group.

If chain has self-transition it is not periodic because it is possible to stay in the same group.

7.1.4 What does the chain do in the long run?

Does $r_{ij}(n) \overset{?}{\to} \pi_j, n \to \infty$, and is it independent from the initial state?

Yes, if:

i.   Recurrent states are all in a single class,
ii. Single recurrent class is not periodic.

Assuming Yes, start from key recursion: $r_{ij}(n) = \sum_k r_{ik}(n-1)p_{kj}$

$$\left. \begin{array}{c} \lim_{n\to\infty} r_{ij}(n) = \pi_j \\ \lim_{n\to\infty} \sum_k r_{ik}(n-1)p_{kj} = \sum_k \pi_k p_{kj} \end{array} \right\} \Rightarrow \pi_j = \sum_k \pi_k p_{kj} - \text{Balance equations.}$$

We get a system of equal unknowns and equations, but one solution is zero → we need one more condition to get the solution to unique solvable linear equations: $\sum_j \pi_j = 1$

7.1.5 Visit frequencies interpretation

We can interpret π_j as a probability or as a frequency at which we find ourselves at state j if we run a long trajectory over that Markov chain.

How often do we get transitions into j? A fraction π_1 of the time we are at 1. Whenever we are at 1, there is going to be a p_{1j} that we make a transition of this kind.

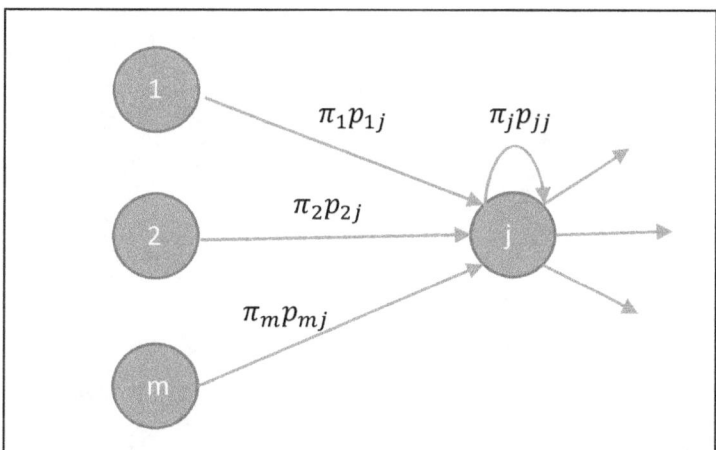

Out of the overall number of transitions that happen at the trajectory, what fraction of those transitions is actually of that kind? That fraction of transitions is the fraction of time that we find ourselves at 1, times the fraction with which out of 1 we happen to find the state j.

- (Long run) frequency of being in j: π_j
- Frequency of transitions $k \rightarrow j$: $\pi_k p_{kj}$
- Frequency of transitions into j: $\sum_k \pi_k p_{kj}$

Probability of state j is the sum of probabilities that the last transition was into state j: $\pi_j = \sum_k \pi_k p_{kj}$

In terms of frequency, the frequency with which we find ourselves at j is the sum of the frequencies of all the possible transition types that take us inside j.

Example: calculate Balance equations

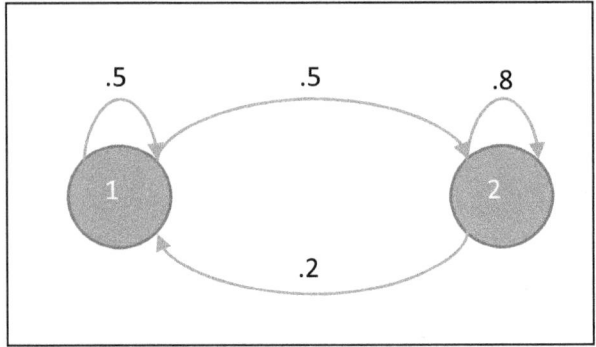

$$\pi_1 = \pi_1 \cdot .5 + \pi_2 \cdot .2$$
$$\pi_2 = \pi_1 \cdot .5 + \pi_2 \cdot .8 \Rightarrow .5\pi_1 = .2\pi_2 \wedge \pi_1 + \pi_2 = 1$$
$$\Rightarrow \pi_1 = \frac{2}{7}, \pi_2 = \frac{5}{7}$$

7.1.6 Birth-Death processes

If we have a large chain it is not a problem for a computer to solve the system. However, we can look for special structures of models that may give more insight or lead us to closed form formulas. Sub-Class of Markov chain in which all these happen is called *Birth-Death processes*. We can go up or down by 1, or stay in place. Example is population: one person dies, one person is born or there is no change.

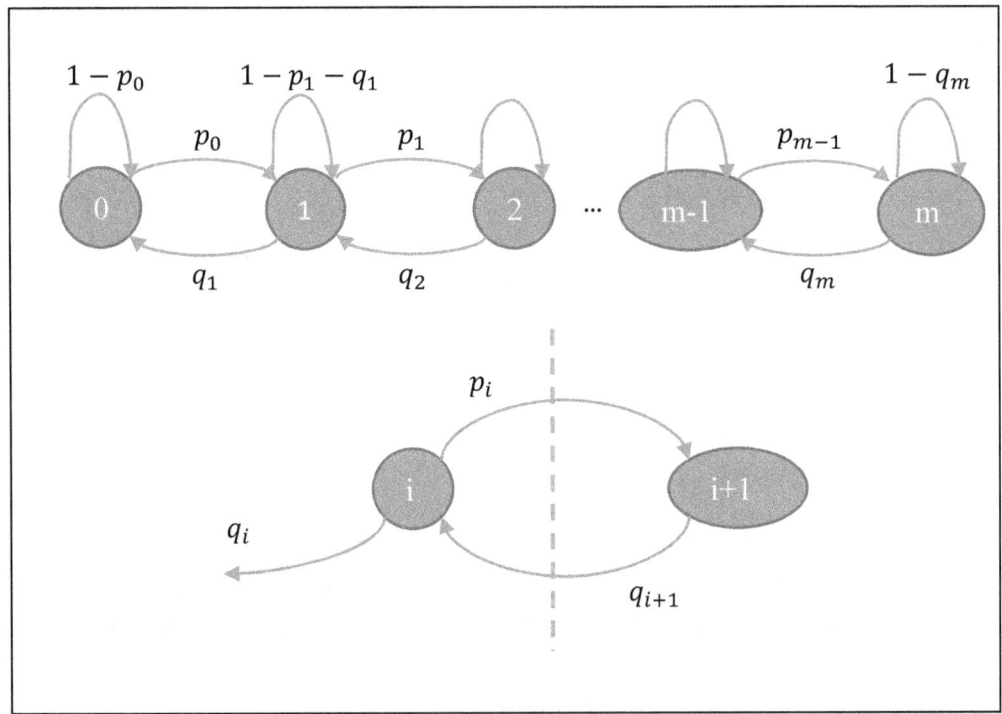

In the long run the frequency of getting forward over the dashed line is the same as going backwards.

We know that it is: $\pi_i p_i = \pi_{i+1} q_{i+1}$, which are recursive equations. We need π_0 in order to solve the system, and we get it from: $\sum \pi_i = 1$.

Special case: $p_i = p, q_i = q$; $\rho = \frac{p}{q}$ – load factor – how loaded the system is: $\rho > 1$ – tendency to the right, $\rho < 1$ – to the left, $\rho = 1$ – equal tendencies.

$$\begin{cases} \pi_{i+1} = \pi_i \rho \Rightarrow \pi_1 = \pi_0 \rho, \pi_2 = \pi_1 \rho = \pi_0 \rho^2, \dots, \pi_i = \pi_0 \rho^i \\ \sum_{i=0}^{m} \pi_0 \rho^i = 1 \Rightarrow \pi_0 = \frac{1}{\sum_{i=0}^{m} \rho^i} \end{cases}$$

Special case: $\rho = 1 \Rightarrow \pi_i = \pi_0$ – every step is equally likely in the long run – symmetric random walk: $\pi_i = \frac{1}{1+m}$

Special case: $m \to \infty, p < q$: $\rho < 1 \Rightarrow \pi_0 = \frac{1}{\sum_{i=0}^{\infty} \rho^i} = \frac{1}{\frac{1}{1-\rho}} = 1 - \rho \Rightarrow \pi_i = (1 - \rho)\rho^i$ – PDF of geometric

distribution shifted to start at 0:

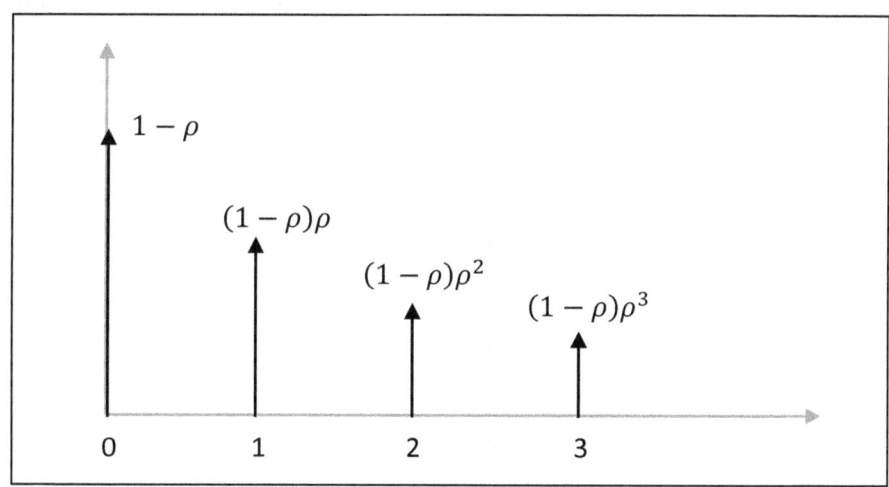

$$E[X_n] = \frac{\rho}{1 - \rho}$$

7.1.7 Multiple recurrent classes

In the case of the Single Recurrent class that is not periodic we have a "Nicest Markov Chain": $\lim_{n \to \infty} r_{ij}(n) = \pi_j$.

With multiple recurrent classes: in the long run there will be transition to some of the recurrent classes. The probabilities of the different states will be the steady state probabilities of the recurrent class chain regarded in isolation. We solve the system of equations just for that chain, and those will be our steady state probabilities.

Example

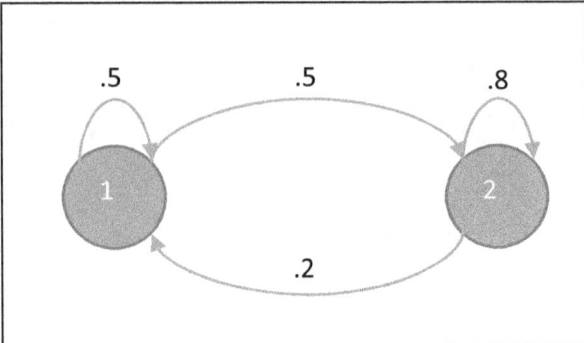

We know that it is: $\pi_1 = \frac{2}{7}, \pi_2 = \frac{5}{7}$. Assume that the process starts at 1: $X_0 = 1$, and calculate:

- $P(X_1 = 1, X_{100} = 1) = P(X_1 = 1 | X_0 = 1)P(X_{100} = 1 | X_1 = 1) = P_{11}r_{11}(99) \approx P_{11}\pi_1$
- $P(X_{100} = 1, X_{101} = 2) = P_{11}(100)P_{12} \approx \pi_1 P_{12}$
- $P(X_{100} = 1, X_{200} = 1) = r_{11}^2(100) \approx \pi_1^2$

We assumed that 99 and 100 is big enough for approximations to be good. This has to do with time scale of Markov Chain – how long does it take for the initial states to be forgotten. In our example, in over 10 steps there is some randomness, in 100 steps there is plenty, so we expect the initial state to be forgotten.

Example

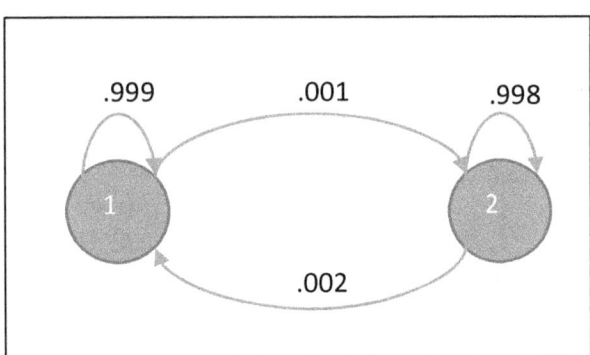

Initial state matters for a long time – we need approximately 10 000 steps before we can start using approximations.

7.1.8 Erlang problem

Erlang was trying to figure out what it would take to properly set up a phone system in a community. The question is how many lines should be set up for a community to communicate to the outside world? We want to set the small number of phone lines to the outside (B), but it has to be large enough so that if reasonable number of people place calls simultaneously, they will all get a line and they will be able to talk → there is a high probability that none would get a busy signal. To set up a model we need two pieces of information:

1. describe how phone calls get initiated,
2. how long does it take until the call is terminated?

The simplest is to assume that the phone calls originate as a Poisson process. Out of the population, people do not coordinate - at random times people will pick up the phone.

Rates that we get from observation:

i. $\lambda \left[\frac{calls}{min}\right]$ – average # of calls per minute,

ii. $\frac{1}{\mu} \left[\frac{min}{call}\right]$ - average call duration.

To get out of continuous time universe we can discretize time with time intervals of small length - δ.

What is the state of the system? We look at the system at some time and see how many lines are busy – that describes the state of the system, so we set up states base: $\{0, \dots, B\}$.

Suppose we have i lines that are busy.

It is possible that the call is terminated and a new call gets placed almost simultaneously → we take δ to be small so that we can ignore this case.

Probabilities:

- Probability of upward transition: probability that Poisson process records an arrival during δ → probability is $\lambda\delta$,
- Probability of Call Termination: call has exponential duration with parameter μ: probability of each is $\mu\delta$ → for i calls the probability for one to terminate is $i\mu\delta$.

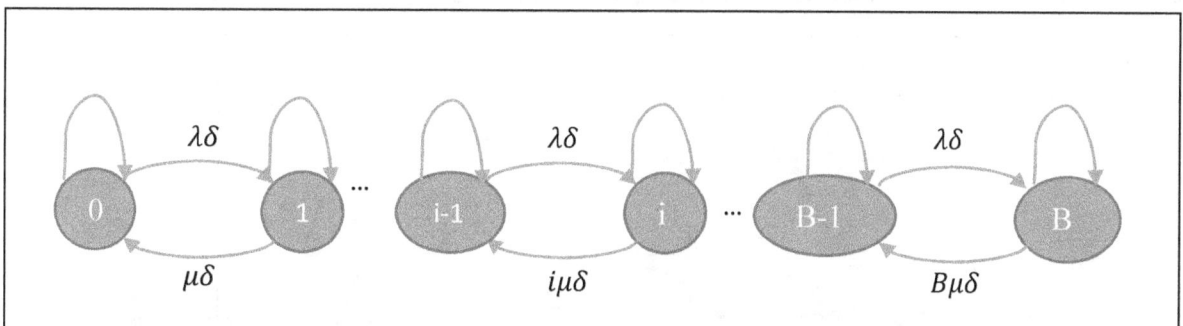

Number of transitions up must be equal to down.

Balance equations: $\lambda\delta \cdot \pi_{i-1} = i\mu\delta \cdot \pi_i \to \lambda \cdot \pi_{i-1} = i\mu \cdot \pi_i \to \pi_i = \frac{\lambda}{i\mu}\pi_{i-1} = \frac{\lambda^i}{\mu^i i!}\pi_0$

$$\sum_{i=0}^{B} \pi_i = 1 \Rightarrow \sum_{i=0}^{B} \frac{\lambda^i}{\mu^i i!}\pi_0 = 1 \Rightarrow \pi_0 = \frac{1}{\sum_{i=0}^{B} \frac{\lambda^i}{\mu^i i!}} = \frac{e^{-\frac{\lambda}{\mu}}\Gamma\left(1+B,\frac{\lambda}{\mu}\right)}{B!}$$

$\pi_B = P(\text{busy})$ – we want it small. How should, given λ and μ, we determine B so that π_B is small?

Example: $\lambda = 30, \mu = \frac{1}{3}, \pi_B \approx 1\%$ → on average 90 calls are active → $B > 90$.

$$\pi_B = \frac{\lambda^B}{\mu^B B!}\pi_0 = \frac{\lambda^B}{\mu^B}\frac{\lambda^B e^{-\frac{\lambda}{\mu}}\Gamma\left(1+B,\frac{\lambda}{\mu}\right)}{(B!)^2} = 1\%$$

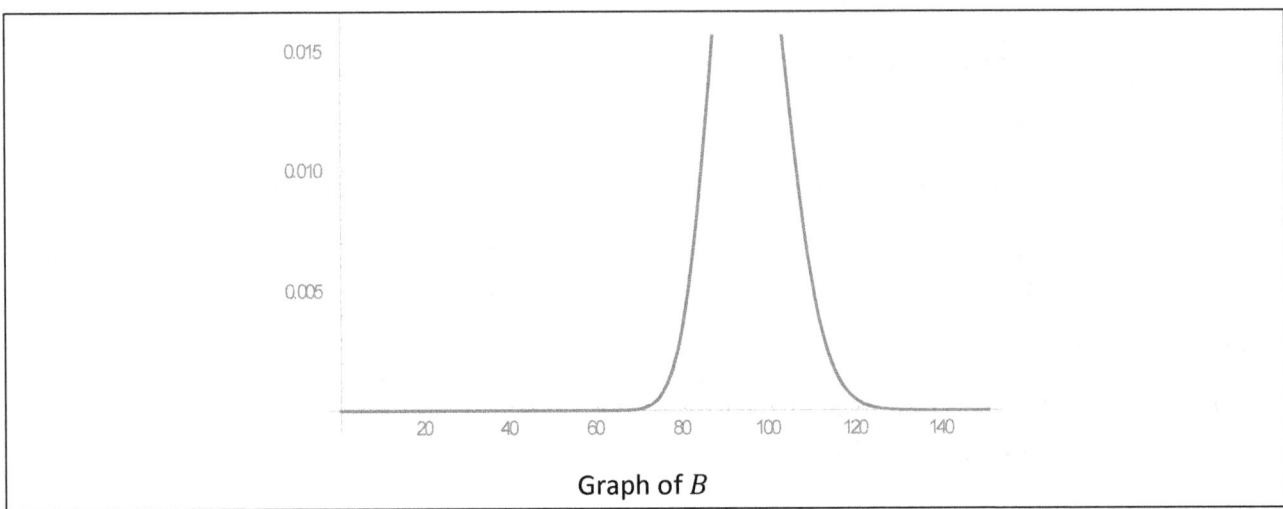

Graph of B

Numerical or graphical solving gives $B = 106$.

7.1.9 Calculating absorption probabilities

Example

What is the probability a_i that process eventually settles in state 4 given that the initial state is i?

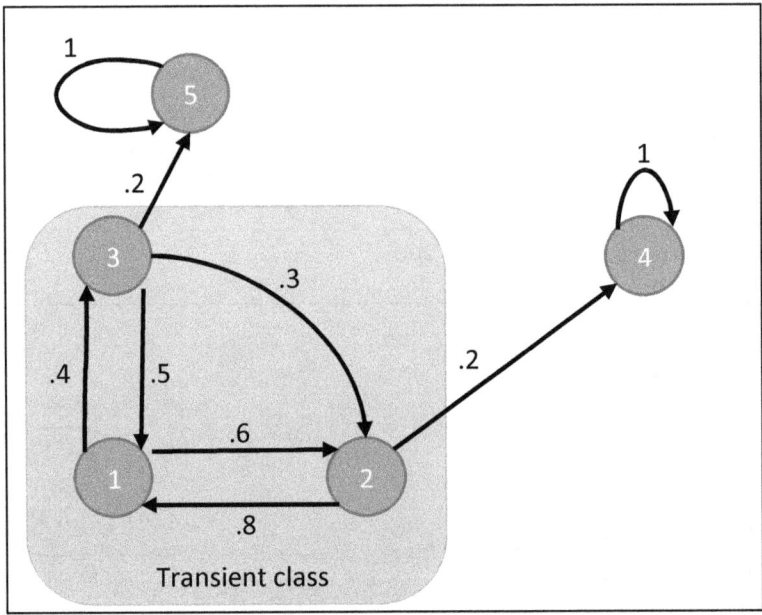

Probability to end up at 4 will depend on the initial state.

Starting at state 2 this can be represented as:

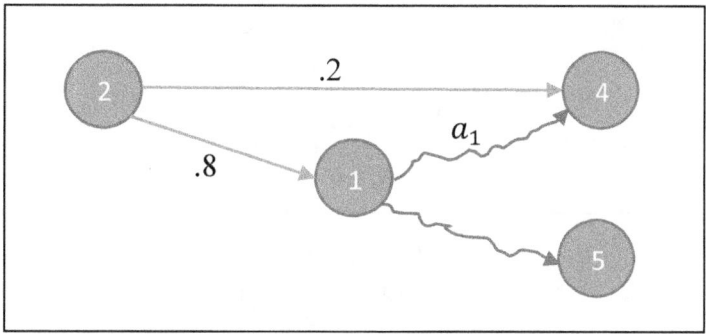

$$a_2 = .2 + .8a_1$$

$(\forall i)\ a_i = \sum_j p_{ij} a_j$ – system of equations.

How do things change if our recurrent, or trapping sets consist of multiple states?

It does not matter if we have multiple states, all that matters is that once we got to recurrent class we are stuck there. So starting from one state we can summarize all the probabilities of going from one state to the states in the class, seeing that class as one entity. If the only thing we care is the probability of ending in the recurrent class, we can replace the class with a single state and calculate probabilities.

How long is it going to take until we get to either state?

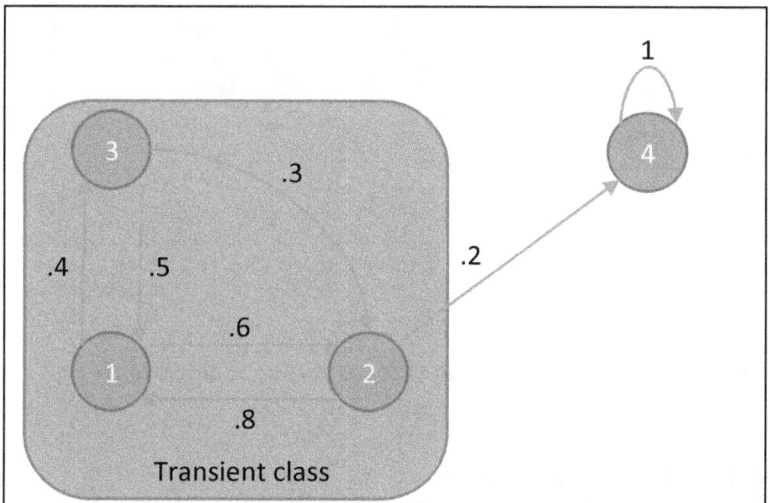

Find expected number of transitions μ_i until we reach the absorbing state, given that initial state is i.

$$\mu_4 = 0$$

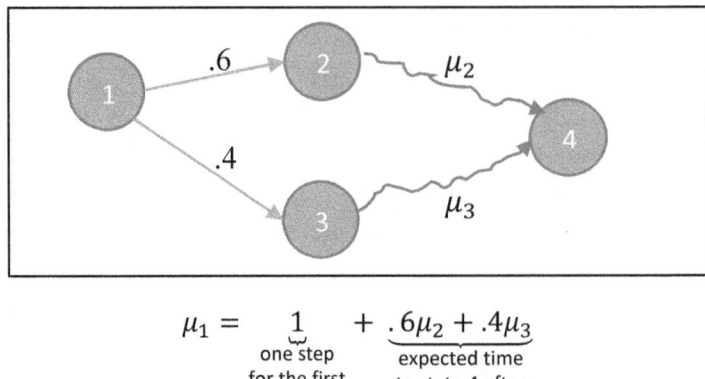

$$\mu_1 = \underbrace{1}_{\substack{\text{one step} \\ \text{for the first} \\ \text{transition}}} + \underbrace{.6\mu_2 + .4\mu_3}_{\substack{\text{expected time} \\ \text{to state 4 after} \\ \text{the first transition}}}$$

$\mu_i = 1 + \sum_j p_{ij}\mu_j$ – system of linear equations of expected times until our chain gets into absorbing state.

Suppose we have transient states and we have multiple absorbing states. We want to calculate expected times until we get to one of the absorbing states. We can group absorbing states and add probabilities for each starting state.

7.1.10 Mean first passage and recurrence times

We have a single recurrent class, and we start from i, and we have special state s. How long is it going to take until we get to s for the first time?

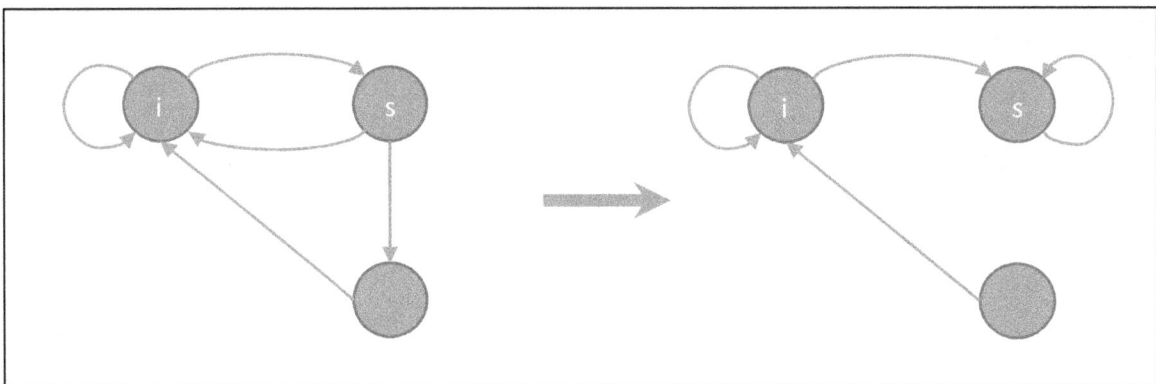

We are not interested in what happens after s → we can change transition out of s to self-transition.

Mean first passage time from i to s: $t_i = E\left[\min\{\ \underbrace{n \geq 0}_{\substack{\text{first time} \\ \text{after everywhere}}}\ , X_n = s\right\}|X_0 = i]$

$\{t_1, \dots, t_m\}$ – unique solutions: $t_s = 0, t_i = 1 + \sum p_{ij}t_j$:

- 1 – one transition,
- $\sum p_{ij}tj$ - after first transition we find ourselves at j with probability p_{ij}, from where we take time t_j to state s.

Mean recurrence time of s

We start at s, how long until we get back at s?

$$t_s^* = E \left[\min\{ \underbrace{n \geq 0}_{\substack{\text{first time} \\ \text{after 0}}}, X_n = s \Big| X_0 = s \right] = 1 + \sum p_{sj} t_j$$

7.2 Programming Markov Models

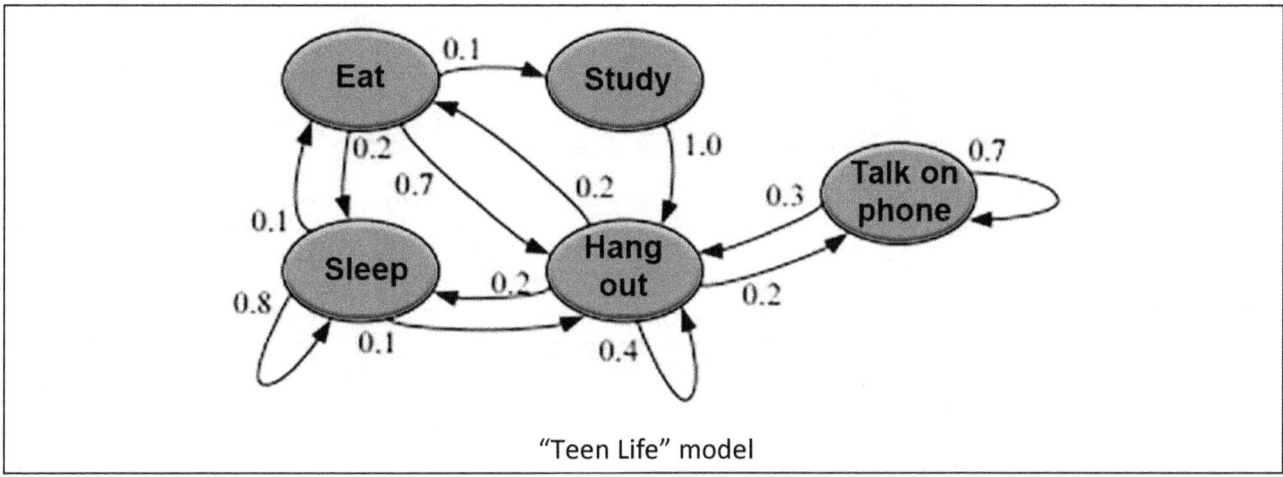

"Teen Life" model

Transition matrix for "Teen Life":

$$A = \begin{pmatrix} \text{in} \downarrow/_{\text{to} \rightarrow} & Eat & Hang & Study & Talk & Sleep \\ Eat & 0 & .7 & .1 & 0 & .2 \\ Hang & .2 & .4 & 0 & .2 & .2 \\ Study & 0 & 1 & 0 & 0 & 0 \\ Talk & 0 & .3 & 0 & .7 & 0 \\ Sleep & .1 & .1 & 0 & 0 & .8 \end{pmatrix}$$

Realization of Markov Model from its $M \times M$ transition matrix using the `Ran`:

`markovgen.h`

```
/* Generate a realization of an M-state Markov model, given its M x M transition matrix
"atrans".
The vector "out" is filled with integers in the range 0...M-1. The starting state is the
optional argument istart (defaults to 0). seed is an optional argument that sets the seed
of the random number generator.
MatDoub is NRmatrix<Doub>;
*/
void markovgen(const MatDoub_I &atrans, VecInt_O &out, Int istart=0, Int seed=1) {
    Int i, ilo, ihi, ii, j, m = atrans.nrows(), n = out.size();
    MatDoub cum(atrans); //Temporary matrix to hold cumulative probabilities
    Doub r;
    Ran ran(seed); //Use Ran
    if (m != atrans.ncols()) throw("transition matrix must be square");
    for (i=0; i<m; i++) { //Fill cum and die if clearly not a transition matrix
        for (j=1; j<m; j++) cum[i][j] += cum[i][j-1];
        if (abs(cum[i][m-1]-1.) > 0.01)
            throw("transition matrix rows must sum to 1");
    }
    j = istart; //The current state is kept in j
    out[0] = j;
```

```
        for (ii=1; ii<n; ii++) { //Main loop.
            r = ran.doub()/cum[j][m-1]; //Slightly-off normalization gets corrected
here
            ilo = 0;
            ihi = m;
            while (ihi-ilo > 1) { //Use bisection to find location among the cumulative
probabilities
                i = (ihi+ilo) >> 1;
                if (r>cum[j][i-1]) ilo = i;
                else ihi = i;
            }
            out[ii] = j = ilo; //Set new current state
    }
}
```

main.cpp

```
// Realization of a Markov Model from M x M transition matrix using Ran
//

#include "stdafx.h"
#include <iostream>
#include "nr3.h"
#include "ran.h"
#include "markovgen.h"
using namespace std;


int main()
{
    MatDoub matrix(5, 5);
    //array for transpose matrix
    double arrayformatrix[5][5] = { {0., .2, 0., 0., .1},
                                    {.7, .4, 1., .3, .1},
                                    {.1, 0., 0., 0., 0.},
                                    {0., .2, 0., .7, 0.},
                                    {.2, .2, 0., 0., .8} };
    for (int i = 0; i < 5; i++)
    {
        for (int j = 0; j < 5; j++)
        {
            matrix[i][j] = arrayformatrix[j][i]; //assign values to matrix, but
it should not be transposed
        }
    }
    VecInt_O vector(5);
    markovgen(matrix, vector, 0);
    cout << "(";
    for (int k = 0; k < 5; k++)
    {
        cout << vector[k];
    }
    cout << ")";
    return 0;
}
```

s_t – population vector whose components give the # of models in each state at t. The components for our model are: {Eat, Hang, Study, Talk, Sleep}.

If all the models in the ensemble are evolved by one transition: $s_{t+1} = A^T s_t$

Evolve n-steps: $s_{t+n} = (A^T)^n s_t$

Every Markov model has at least one equilibrium distribution of states that remains unaffected when multiplied by A^T: $A^T s_e = s_e \Leftrightarrow (A^T - \mathbb{I})s_e = 0 \rightarrow A^T$ has at least one eigenvalue that is equal to 1: $(\exists j)\lambda_j = 1$ and for that eigenvalue the corresponding eigenvector is s_e.

Δ.

$$A^T = \begin{pmatrix} 0 & 0.2 & 0 & 0 & 0.1 \\ 0.7 & 0.4 & 1 & 0.3 & 0.1 \\ 0.1 & 0 & 0 & 0 & 0 \\ 0 & 0.2 & 0 & 0.7 & 0 \\ 0.2 & 0.2 & 0 & 0 & 0.8 \end{pmatrix}$$

$$\det(A^T - \lambda\mathbb{I}_{5\times5}) = 0 \Rightarrow 0.0098 + 0.0422\lambda - 0.032\lambda^2 - 0.92\lambda^3 + 1.9\lambda^4 - \lambda^5 = 0$$

$$\lambda = \{-0.139 - 0.107i, -0.139 + 0.107i, 0.416, 0.762, 1\}$$

$$(A^T - \mathbb{I})s_e = 0 \Rightarrow (1, 3, 0.1, 2, 4)$$

We need to normalize this vector to get s_e (normalize so that the sum of the components is equal to 1):

$$s_e = \frac{(1, 3, .1, 2, 4)}{1 + 3 + .1 + 2 + 4} \approx (0.1, 0.3, 0.01, 0.2, 0.4) \blacksquare$$

If starting distribution converges to a unique equilibrium → passes the tests of *irreducibility* and *aperiodicity* → ergodic.

It does not converge to a unique equilibrium:

- If it has more than one eigenvalue equal to 1 → model will converge to some different linear combination of the corresponding eigenvectors for different starting distributions – fails the test of *irreducibility*.
- If it has a periodic limit cycle so that, for most starting distributions, it doesn't converge at all – fails the test of *aperiodicity*.

Ergodic test: perform successive squaring of the A^T to take it to a very high power of 2^{32}: $(A^T)^{2^{32}}$ → if all columns are converging to identical vectors → there is just one eigenvalue equal to 1 → all starting distributions will converge to its eigenvector (repeated column vector).

Example: multiple equilibria and periodic limit cycles:

$$A = \begin{pmatrix} 0 & 1 & 0 & 0 & 0 \\ 1 & 0 & 0 & 0 & 0 \\ 0 & 0.7 & 0 & 0.3 & 0 \\ 0 & 0 & 0 & 0 & 1 \\ 0 & 0 & 0 & 1 & 0 \end{pmatrix}$$

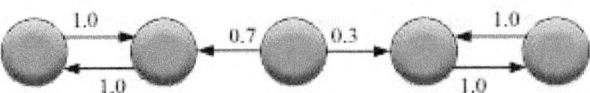

We can guess the eigenvectors of A^T with unit eigenvalue, since it will be stuck to the two left states or to the two right states: $(.5, .5, 0, 0, 0)$ and $(0, 0, 0, .5, .5)$.

$$(A^T)^{2^{32}} = \begin{pmatrix} 1 & 0 & .7 & 0 & 0 \\ 0 & 1 & 0 & 0 & 0 \\ 0 & 0 & 0 & 0 & 0 \\ 0 & 0 & 0 & 1 & 0 \\ 0 & 0 & .3 & 0 & 1 \end{pmatrix}$$ – the matrix does not have any of these eigenvectors as its columns →

the model has only unstable equilibria. We can spot two identity matrices in it: green and blue – these blocks are representing the two limit cycles (which have period 2, so the dimension of the block is 2).

If we know of guess that there is a single stable equilibrium → successive squaring is a poor way to get the equilibrium distribution. A better way is inverse iteration – solve the equation by LU decomposition[8]:

$$(A^T - .999999\mathbb{I})s_e = \begin{pmatrix} 1 \\ \vdots \\ 1 \end{pmatrix}$$

7.3 Hidden Markov Models HMM[9]

We now turn to the real business, which is to estimate statistically the state of a HMM, given only partial or imperfect information.

HMM is a statistical Markov model in which the system being modeled is assumed to be a Markov process with unobserved (hidden) states.

In simpler Markov models, like a Markov chain, the state is directly visible → the state transition probabilities are the only parameters.

In a HMM, the state is not directly visible, but the output (which depends on the state) is. Each state has a probability distribution over the possible output tokens → the sequence of tokens gives some information about the sequence of states. The adjective "hidden" refers to the state sequence through which the model passes, not the parameters of the model.

---

[8] It is a simple but effective method described in NR, section 2.3.
[9] The applications of HMM in physics are through Mori and Zwanzig formalism, which is out of the scope of this book. Most applications of HMM are in machine learning.

Example: Balls in Bowls

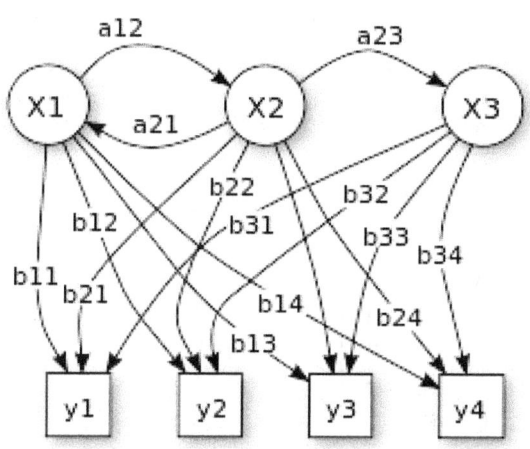

Rules:

- X1, X2 and X3 are the bowls (states), each contains a balls labeled y1, y2, y3, y4 (symbols – outputs).
- The balls are drawn by the machine, and we can only see the sequence of balls that are drawn – we do not know from which bowl.
- The machine works so that the choice of the bowl for the nth ball depends only upon a random number and the choice of the bowl for the $(n-1)^{th}$ bowl – Markov process.
- The Markov process itself cannot be observed – that's why it's called "hidden".
- We cannot be sure from which bowl did the particular ball came, but we can work out the probabilities.
- We get the output in the form of a sequence, for example: $\{y3, y4, y1, y1, y2, y3, ..., y_1\}$.
- The choice of the bowls is dictated by the state transition matrix.

Example: "Teen Life"

The teen is locked up in the room. We can only interact with the teen by talking through the closed door. We know that the states are: Eat, Hang, Study, Talk and Sleep. The observations we can get are: silence, "I'm studying", "I'm busy", grunt or snore.

The observations – emitted symbols – give only incomplete state information: e.g. if the response is "I'm studying", the teen might be lying and instead be talking on the phone, as we can see from the table.

			$i = 0$	1	2	3	4
k	symbol	meaning	Eat	Hang	Study	Talk	Sleep
0	o	Silence	.2	.2	0	.3	.5
1	s	"I'm studying"	0	0	1	.2	0
2	b	"I'm busy"	0	.6	0	.4	0
3	g	Grunt	.8	.2	0	.1	0
4	z	Snore	0	0	0	0	.5

A state can give more than one possible symbol, and a symbol can be emitted by more than one possible state. Our goal is to make the best statistical reconstruction of *s* from *y*.

What data do we have?

i. We have a transition matrix for "teen life":

$$A = \begin{pmatrix} 0 & .7 & .1 & 0 & .2 \\ .2 & .4 & 0 & .2 & .2 \\ 0 & 1 & 0 & 0 & 0 \\ 0 & .3 & 0 & .7 & 0 \\ .1 & .1 & 0 & 0 & .8 \end{pmatrix}$$

ii. Emission probabilities from the table above:

$$B = \begin{pmatrix} .2 & .2 & 0 & .3 & .5 \\ 0 & 0 & 1 & .2 & 0 \\ 0 & .6 & 0 & .4 & 0 \\ .8 & .2 & 0 & .1 & 0 \\ 0 & 0 & 0 & 0 & .5 \end{pmatrix}$$

iii. Vector labeling symbols, from the table:

$$(o \quad s \quad b \quad g \quad z)$$

iv. Vector consisting of the sequence of observations. For example, we can say that it is:

$$y = (s \quad b \quad o \quad z \quad g \quad g \quad s \quad o \quad b \quad s \quad s)$$

HMM model is characterized by:

1. M - # of states of the model: $s_i \in [0, M-1]$. We label the state at time t: s_t, and the model evolves through N timesteps: $t \in [0, N-1] \rightarrow$ and we can generate a sequence:
$$s = \{s_t\} = (s_0, \dots, s_{N-1}) \text{ each in the range } s_i \in [0, M-1]^{10}$$

2. K - # (set) of distinct observation symbols. We label symbols with $k \in [0, K-1]$, chosen probabilistically from K. The observations are a vector of integers:
$$y = \{y_t\} = (y_0, \dots, y_{N-1}), y_i \in [0, K-1]$$

3. State transition probability distribution (matrix) A, with elements $A_{ij} = P(s_j|s_i)$ – transition probability from the state i to state j. If state i cannot directly reach state j (i.e. in one step):
$A_{ij} = 0$

4. The probability of getting k given i: $b_i(k) \equiv P(k|i), i \in [0, M-1], k \in [0, K-1]$
Normalization condition: $\sum_{k=0}^{K-1} b_i(k) = 1$
Observation symbol probability distribution in state i:
$$B = \{b_i(k)\}_{K \times M} = \begin{pmatrix} b_0(0) & \cdots & b_{M-1}(0) \\ \vdots & \ddots & \vdots \\ b_0(K-1) & \cdots & b_{M-1}(K-1) \end{pmatrix}$$

5. The initial state distribution: $\pi(s_{t=0})$

---

[10] Note: watch that NR gives the subscript t and i for s, which are different things. One could say that the proper labeling would be: s_{it}.

Given appropriate values of M, N, A, B, π, the HMM can be used as a generator to give y, as follows:

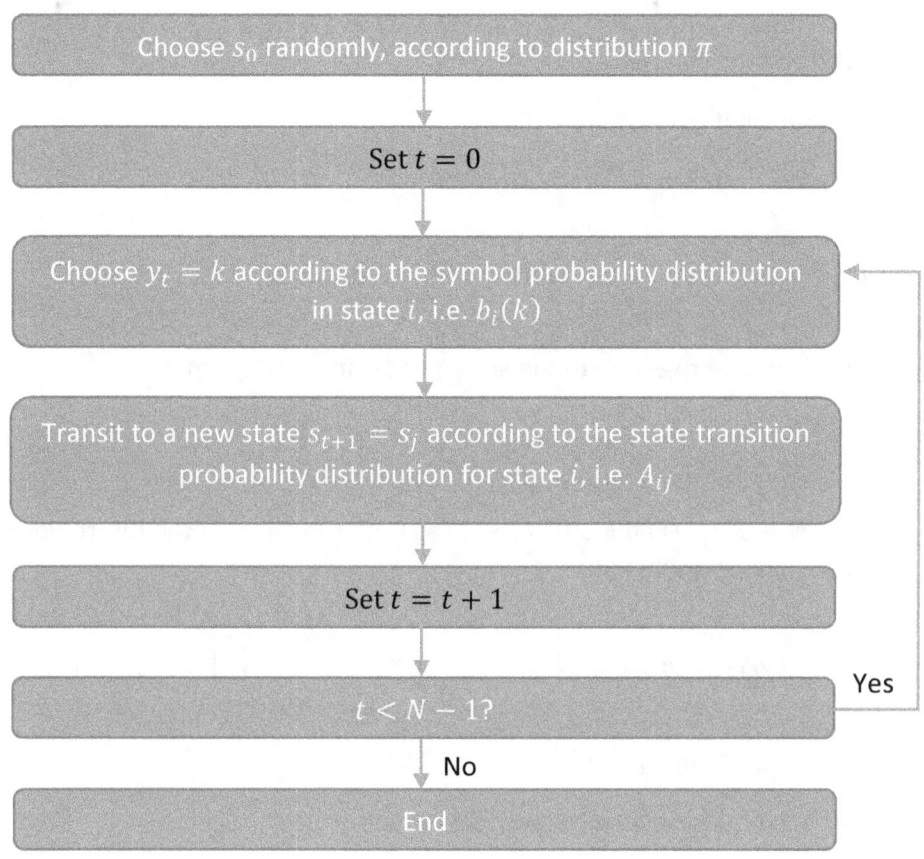

The complete specification of HMM requires specification of observations symbols, and the specification of three probability measures: $\lambda = (A, B, \pi)$

7.3.1 The Tree Basic Problems for HMM

There are three basic problems of interest that must be solved for the model to be useful in real-world applications:

1. Given the y and λ, how do we efficiently compute $P(y|\lambda)$?
2. Given the y and λ, how do we chose s that best explains the observations? More specifically, at each t we want to estimate the probability that the actual state is i, given all the data:

$$P_t(i) \equiv P(s_t = i|y)$$

3. How do we adjust the λ to maximize $P(y|\lambda)$?

Problem 1

How do we compute the probability that the observed sequence was produced by the model – scoring how well a given model matches y → allows us to choose the model which best matches the observations.

The most straightforward way is through enumerating every possible state sequence of length N.

$$P(y|s,\lambda) = \prod_{t=0}^{N-1} P(y_t|s_t,\lambda) = b_{i_0}(y_0) \dots b_{i_{N-1}}(y_{N-1})$$

Forward estimate

$\alpha_t(i)$ – probability of the observed data until t, given that we are in i at t (that is, $y_0, \dots, y_t$); can also be interpreted as the likelihood of the state, given the data.

We must sum over all possible paths that get to state i at t:

$\alpha_0(i) = b_i(y_0)$ – the probability at $t = 0$ that we get y_0 from state i,

$$\alpha_t(i) = \sum_{i_0,\dots,i_{t-1}} b_{i_0}(y_0)A_{i_0 i_1}b_{i_1}(y_1)\dots A_{i_{t-1}i}b_i(y_t) = \sum_{i_0,\dots,i_{t-1}} b_{i_0}(y_0)\prod_{j=1}^{t} A_{i_{j-1}i_j}b_{i_j}(y_j)$$

The equation uses the data earlier in time to estimate the state i at t – forward estimate.

Recurrence relation: $\alpha_{t+1}(j) = \sum_{i=0}^{M-1} \alpha_t(i)A_{ij}b_j(y_{t+1}), j \in [0,\dots,M-1], t = 0,\dots,N-2$

Backward estimate

We now define the $\beta_t(i)$ for $t = N-1,\dots,0$ and $i \in [0, M-1]$ as the probability of the future observed data $(y_{t+1}, y_{t+2}, \dots, y_{N-1})$, given that we are in i at t:

$$\beta_t(i) = \sum_{i_{t+1},\dots,i_{N-1}} A_{i i_{t+1}}b_{i_{t+1}}(y_{t+1})\dots A_{i_{N-2}i_{N-1}}b_{i_{N-1}}(y_{N-1}) = \sum_{i_{t+1},\dots,i_{N-1}} \prod_{j=t+1}^{N-1} A_{i_{j-1}i_j}b_{i_j}(y_j)$$

Backward recurrence: $\beta_{t-1}(i) = \sum_{j=0}^{M-1} A_{ij}b_j(y_t)\beta_t(j), t = N-1,\dots,1$

Calculating all the β's for $t = N-1,\dots,0$ is called a backward estimate.

Problem 2

Attempt to uncover the hidden part of the model, i.e. to find the optimal state sequences, or to get average statistics of individual states.

Unlike problem 1 for which an exact solution can be given, there are several possible ways of solving problem 2.

The payoff of forward-backward estimates

* $\alpha_t(i)\beta_t(i)$ – unnormalized probability of i at t given all the data.
* Normalization constant: $\mathcal{L} = \sum_{i=0}^{M-1} \alpha_t(i)\beta_t(i)$ – independent of t so we can calculate it just once.
* Probability of each separate state at each separate time: $P_t(i) = \frac{\alpha_t(i)\beta_t(i)}{\mathcal{L}}, \sum_t P_t(i) = 1$

Now we can solve for the individually most likely state $s_t = \underset{i\in[0,M-1]}{\mathrm{argmax}}\, P_t(i)$[11]

---

[11] In contrast to global maxima, referring to the largest *outputs* of a function, argmax refers to the *inputs*, or arguments.

Problem 3

We attempt to optimize λ to best describe $y \rightarrow y$ is a training sequence, since it is used to train the HMM.

This can also be labeled as Bayesian Re-Estimation of the Model Parameters.

The probability that we were in i at t is $P_t(i)$. What is the probability, given y, that a given transition, between t and $t + 1$, was a transition between i and j?

$$P(s_t = i, s_{t+1} = j | y) = \frac{\underbrace{P(y|s_{t+1} = j, s_t = i)}_{\alpha_t(i) b_j(y_{t+1}) \beta_{t+1}(j)} \underbrace{P(s_{t+1} = j | s_t = i)}_{A_{ij}}}{\sum_j P(y|s_{t+1} = j, s_t = i) P(s_{t+1} = j | s_t = i)} \underbrace{\frac{P(s_t = i | y)}{P_t(i)}}$$

$$= \frac{\alpha_t(i) b_j(y_{t+1}) \beta_{t+1}(j) A_{ij}}{\cancel{\alpha_t(i)} \underbrace{\sum_j b_j(y_{t+1}) \beta_{t+1}(j) A_{ij}}_{\beta_t(i)}} \cdot \frac{\cancel{\alpha_t(i)}\cancel{\beta_t(i)}}{\mathcal{L}} = \frac{\alpha_t(i) A_{ij} b_j(y_{t+1}) \beta_{t+1}(j)}{\mathcal{L}}$$

For a long run of data, we can compute the fraction of the time that a state i transitions to state j as the estimated number of $i \rightarrow j$ transitions divided by the estimated number of i states (Baum-Welch re-estimation equations):

- $\hat{A}_{ij} = \frac{\sum_t \alpha_t(i) A_{ij} b_j(y_{t+1}) \beta_{t+1}(j)}{\sum_t \alpha_t(i) \beta_t(i)}$ – a re-estimation of A_{ij}
- $\hat{b}_i(k) = \frac{\sum_t \delta(y_t, k) \alpha_t(i) \beta_t(i)}{\sum_t \alpha_t(i) \beta_t(i)}$ – re-estimation of $b_i(k)$
- $\delta(j, k) = \begin{cases} 1, & j = k \\ 0, & \text{otherwise} \end{cases}$

Baum and Welch showed that replacing matrices by their re-estimates, and then recalculating the probabilities of each state at each time by the forward-backward algorithm, always increases $\mathcal{L}$, the overall likelihood of the model. We can continue this cycle of re-estimating model probabilities (Baum-Welch), until convergence to a max is achieved.

We began by estimating states in a known HMM, and just from the data, we can get estimate of the states and estimate of the model itself – transition and symbol probabilities. Like any iterative process, this works best if we have a good initial guess, but it will often converge to a good model from a fairly random initial guess.

7.3.2 Code

We construct a HMM structure by specifying:

i. transition probability matrix A,
ii. symbol probability matrix $b_{ik} \equiv b_i(k)$,
iii. vector of observed data y.

Matrices are passed in a form of `MatDoub`, vector in `VecInt`.

To do the forward-backward estimation, we call the function `forwardbackward`, which fills the matrix $pstate_{ti} = P_t(i)$. It also sets the internal variables `lhood` and `lrnrm` so that the function `loglikelihood` returns the log $\mathcal{L}$.

BIG, BIGI, `arnrm`, `brnrm` and `lrnrm` all relate to dealing with values that would underflow an ordinary floating format.

At the end, when α, β and $\mathcal{L}$ are combined, we get probability values with reasonable magnitues.

For our example `forwardbackward`, we get the hidden states of "Teen Life", given just a long string of output symbols. If we take the prediction to be the state with the highest probability at each time, then this is correct about 78% of the time, for 17% of the cases it is the second-highest probability.

The output is a prediction and a quantitative assessment of how sure the model is of that prediction.

We must always precede a call to `baumwelch` by a call to `forwardbackward`, since it updates α and β tables.

`hmm.h`

```
struct HMM { //Structure for a HMM
        MatDoub a, b; //a - transition matrix, b - symbol probability matrix
        VecInt obs; //Observed data
        Int fbdone;
        Int mstat, nobs, ksym; //Number of states, observations and symbols
        Int lrnrm;
        MatDoub alpha, beta, pstate; //Matrices alpha, beta and P_i(t)
        VecInt arnrm, brnrm;
        Doub BIG, BIGI, lhood;
        HMM(MatDoub_I &aa, MatDoub_I &bb, VecInt_I &obs); //Constructor
        void forwardbackward(); //HMM state estimation
        void baumwelch(); //HMM parameter re-estimation
        Doub loglikelihood() {return log(lhood)+lrnrm*log(BIGI);} //Returns the log-
likelihood computed by forwardbackward()
};

HMM::HMM(MatDoub_I &aa, MatDoub_I &bb, VecInt_I &obss) :
        a(aa), b(bb), obs(obss), fbdone(0),
        mstat(a.nrows()), nobs(obs.size()), ksym(b.ncols()),
        alpha(nobs,mstat), beta(nobs,mstat), pstate(nobs,mstat),
        arnrm(nobs), brnrm(nobs), BIG(1.e20), BIGI(1./BIG)  {
        /*Constructor. Inputs are:
                *transition matrix aa,
                *symbol probability matrix bb,
                *observed vector of symbols obss.
           Local copies are made, so the input quantities need not be preserved by the
calling program.
        */
        Int i,j,k;
        Doub sum;
        //Code for checking input
                                               .
                                               .
                                               .

void HMM::forwardbackward() { //forward-backward algorithm. Using the stored a, b, and
obs matrices, the matrices alpha, beta and pstate are calculated. The latter is the state
estimation of the model, given the data.
        Int i,j,t;
        Doub sum,asum,bsum;
        for (i=0; i<mstat; i++) alpha[0][i] = b[i][obs[0]];
```

```
        arnrm[0] = 0;
        for (t=1; t<nobs; t++) { //Forward pass
                    .
                    .
                    .
                arnrm[t] = arnrm[t-1]; //Renormalize the alphas as necessary to avoid
underflow, keeping track of how many renormalizations for each alpha
                    .
                    .
                    .
        for (t=nobs-2; t>=0; t--) { //Backward pass
            bsum = 0.;
                    .
                    .
                    .
            if (bsum < BIGI) { //Renormalize, like alpha, if necessary
                    ++brnrm[t];
                    for (j=0; j<mstat; j++) beta[t][j] *= BIG;
            }
        }
        lhood = 0.; //Overall likelihood is lhood with lnorm renormalizations
        for (i=0; i<mstat; i++) lhood += alpha[0][i]*beta[0][i];
        lrnrm = arnrm[0] + brnrm[0];
        while (lhood < BIGI) {lhood *= BIG; lrnrm++;}
        for (t=0; t<nobs; t++) { //Get state probabilities from alphas and betas
            sum = 0.;
            for (i=0; i<mstat; i++) sum += (pstate[t][i] = alpha[t][i]*beta[t][i]);
            // sum = lhood*pow(BIGI, lrnrm - arnrm[t] - brnrm[t]);
            for (i=0; i<mstat; i++) pstate[t][i] /= sum;
        }
        fbdone = 1; // Flag prevents misuse of baumwelch(), later.
}
void HMM::baumwelch() { //Baum-Welch re-estimation of the stored matrices a and b, using
the data obs and the matrices alpha and beta as computed by forwardbackward() (which must
be called first). The previous values of a and b are overwritten.
                    .
                    .
                    .
        for (j=0; j<mstat; j++) { //Inner loop over j gets elements of a
                    .
                    .
                    .
        fbdone = 0; // stops the routine to be called again until forwardbackward() has
benn called
}
```

8. Markov Chain Monte Carlo (MCMC)

MCMC methods are a class of algorithms for random sampling from a probability distribution based on constructing a Markov chain that has the desired distribution as its equilibrium distribution. Unlike MC, the goal is not to sample a multidimensional region uniformly, but to visit a point x with a probability proportional to some given function $\pi(x)$ proportional to distribution function (it might not be normalized).

The state of the chain after a number of steps is then used as a sample of the desired distribution.

- x – point, values of the model parameters,
- D – set of data,
- $P(D|x)$ – probability of D given x.

We assume $P(x) \to \pi(x) = P(D|x)P(x)$ – proportional to probability of the model.

We could obtain all the same information by MC, computing $\pi(x_i)$, but MCMC puts its sample points where $\pi(x)$ is large, which is a huge advantage in high dimensional space or where $\pi(x)$ is expensive to compute.

MCMC: sample $\pi(x)$ via Markov Chain, a sequence of points $\{x_0, ...\}$ that are correlated and due to the Ergodic theorem, visit every point x in proportion to $\pi(x)$.

Markov property: x_i is chosen from a distribution that depends only on the value of the immediately preceding point x_{i-1} – the chain has the memory extending only to one previous point and is completely defined by a transition probability function $p(x_i|x_{i-1})$. If transition probability function is chosen to satisfy the *detailed balance equation:* $\pi(x_1)p(x_2|x_1) = \pi(x_2)p(x_1|x_2)$ then the Markov Chain will sample $\pi(x)$ ergodically.

$x_1 \propto \pi(x_1) \wedge x_2 \propto \pi(x_2) \to$ overall transition rates in each direction (the product of a population density and a transition probability) are the same, and we write: $x_1 \leftrightarrow x_2$.

$\Delta. \pi(x_1)p(x_2|x_1) = \pi(x_2)p(x_1|x_2) \to \int \pi(x_1)p(x_2|x_1)dx_1 = \pi(x_2) \underbrace{\int p(x_1|x_2)dx_1}_{1}$

Left side is the probability of x_2 computed by integrating over all possible values of x_1 with the corresponding transition probability. The last equation says that if $x_1 \sim \pi \to x_2 \sim \pi$ ∎.

If x_i are discrete $\to p(x_j|x_i) \equiv P_{ij}$ is a transition matrix.

π is an eigenvector with unity eigenvalue: $P\pi(x) = \pi(x)$

8.1 Metropolis-Hastings Algorithm

The Metropolis algorithm is used in the importance sampling of multidimensional integrals. It generates a random walk of points distributed according to a required probability distribution. From an initial position in phase or configuration space, a proposed move is generated and the move either accepted or rejected according to the Metropolis algorithm. By taking a sufficient # of trial steps all of phase space is explored and the Metropolis algorithm ensures that the points are distributed according to the required probability distribution.

In other words: the Metropolis algorithm is a MCMC method for obtaining a sequence of random samples from a probability distribution for which direct sampling is difficult. The method generates a random walk using a proposal density and a method for rejecting some of the proposed moves.

If we can find $p(x_2|x_1)$ that satisfies the detailed balance equation, we can use this algorithm:

- Pick a *proposal distribution* $q(x_2|x_1)$ as long as it is Ergodic. It suggests a candidate for the next sample value x_2 given previous sample value x_1. It must be symmetric: $q(x_2|x_1) = q(x_1|x_2)$. A usual choice is to let q be a Gaussian distribution centered at x_1 so that the points closer to it are more likely to be visited next — making the sequence of samples into a random walk.
- We want to generate a step starting at x_1 (and repeat for each iteration):
  - First generate a candidate point $x_{2c} \sim q(x_2|x_1)$,
  - Second, calculate an acceptance probability: $\alpha(x_1, x_{2c}) = \min\left(1, \frac{\pi(x_{2c})q(x_1|x_{2c})}{\pi(x_1)q(x_{2c}|x_1)}\right)$, which is used to decide whether to accept or reject the candidate.
  - With probability $\alpha(x_1, x_{2c})$ we accept the candidate point and set $x_2 = x_{2c}$; otherwise reject it and leave the point unchanged: $x_2 = x_1$.
  - The net result of this process is a transition probability:
  $$p(x_2|x_1) = q(x_2|x_1)\alpha(x_1, x_2), x_2 \neq x_1$$

The algorithm proceeds by randomly attempting to move about the sample space, sometimes accepting the moves and sometimes remaining in place. α indicates how probable the new proposed sample is, with respect to the current sample. If we attempt to move to a point that is more probable than the existing point (i.e. a point in a higher-density region of $p(x)$) we will always accept the move. If we attempt to move to a less probable point, we will sometimes reject the move. We will tend to stay in (and return large number of samples from) high density regions of $p(x)$, while only occasionally visiting low-density regions — this is why this algorithm works, and returns samples that follow the $p(x)$.

It is possible to choose the proposal distribution in such a way as to simplify
$$\alpha(x_1, x_{2c}) = \min\left(1, \frac{\pi(x_{2c})q(x_1|x_{2c})}{\pi(x_1)q(x_{2c}|x_1)}\right)$$

8.1.1 A Worked Example
At the beginning of an experiment, events occur Poisson randomly with a mean rate λ_1, but only every k_1^{th} event is recorded. Then, at time t_c, the mean rate changes to λ_2, but now every k_2^{th} event is recorded. We are given times $t_0, \dots, t_{N-1}$ of the N recorded events. $\lambda_1, \lambda_2, k_1, k_2, t_c$ — unknowns we want to find.

To summarize:

- Poisson process,
- State (point) x is defined by (λ, k, t),
- $(\lambda_1, k_1) \overset{t_c}{\to} (\lambda_2, k_2)$,
- N — number of recorded events,
- $\{t_0, \dots, t_{N-1}\}$ — times of events being recorded, t_i - i^{th} recorded time.
- $\{\lambda_1, \lambda_2, k_1, k_2, t_c\}$ =?

Parts 1-3 are all problem dependent and benefit from cleverness and special tricks:

Part1

First we need an object that represents x. For that we create the blueprints-structure.

mcmc.h

```
struct State { //Structure containing the components of x
        Doub lam1, lam2; // lambda_1 and lambda_2
        Doub tc; //t_c
        Int k1, k2; //k_1 and k_2
        Doub plog; // log(P) - not a part of x

        State(Doub la1, Doub la2, Doub t, Int kk1, Int kk2) : //The constructor
                lam1(la1), lam2(la2), tc(t), k1(kk1), k2(kk2) {} //Set the initial values
        State() {};
};
```

Part2

Next, we need an object for calculating $\pi(x) = P(D|x)$ – the probability of the data given the parameters. If a Poisson process has a rate $\lambda \rightarrow$ the waiting time to the kth event is distributed as $\text{Gamma}(k, \lambda)$.

Let's first define time interval: $\tau = t_{i+k} - t_i$ – time interval between two events, in general. Using Markov property, we are interested only in interval that depends only from the previous interval: $\tau = t_{i+1} - t_i$

Properties of Gamma distribution: $x \sim \text{Gamma}(\alpha, \beta), \alpha > 0, \beta > 0; p(x) = \frac{\beta^\alpha}{\Gamma(\alpha)} x^{\alpha-1} e^{-\beta x}, x > 0 \rightarrow |x \equiv$

$\tau, \alpha \equiv k, \beta \equiv \lambda| \rightarrow p(\tau|k, \lambda) = \frac{\lambda^k}{\Gamma(k)} \tau^{k-1} e^{-\lambda \tau} = \frac{\lambda^k}{(k-1)!} \tau^{k-1} e^{-\lambda \tau}$

Probabilities for non-overlapping intervals are independent $\rightarrow$ we can separate probabilities for each interval $\wedge$ after t_c the parameters change $\rightarrow$

$$P(D|x) = \prod_{t_i \leq t_c} p(\tau|k_1, \lambda_1) \cdot \prod_{t_i > t_c} p(\tau|k_2, \lambda_2)$$

As the amount of data gets large it is better to calculate $\log P$ instead.

$$\log p(\tau|k, \lambda) = \log\left(\frac{\lambda^k}{(k-1)!} \tau^{k-1} e^{-\lambda \tau}\right) =$$

$$\log P(D|x) = \log\left(\prod_{t_i \leq t_c} p(\tau|k_1, \lambda_1) \cdot \prod_{t_i > t_c} p(\tau|k_2, \lambda_2)\right) = \log \prod_{t_i \leq t_c} p(\tau|k_1, \lambda_1) + \log \prod_{t_i > t_c} p(\tau|k_2, \lambda_2)$$

$$= \sum_{t_i \leq t_c} \log(p(\tau|k_1, \lambda_1)) + \sum_{t_i > t_c} \log(p(\tau|k_2, \lambda_2))$$

It is important to make the calculation as efficient as possible, because it will be done at every step. Particularly important is to minimize the amount of looping over all the data points. An efficient way to proceed is to digest the data once and store two cumulative sums. Then, given t_c, we can read left and right sums $\rightarrow$ there is no loop over data. The resulting object:

mcmc.h

```
struct Plog { //functor that calculates logP of a State
      VecDoub &dat; //Bind to data vector
      Int ndat;
      VecDoub stau, slogtau;

      Plog(VecDoub &data) : dat(data), ndat(data.size()),
      stau(ndat), slogtau(ndat) { //Constructor. Digest the data vector for subsequent
fast calculation of logP. The data are assumed to be sorted in ascending order.
            Int i;
            stau[0] = slogtau[0] = 0.;
            for (i=1;i<ndat;i++) {
                  stau[i] = dat[i]-dat[0]; //Equal to sum of intervals
                  slogtau[i] = slogtau[i-1] + log(dat[i]-dat[i-1]);
            }
      }

      Doub operator() (State &s) { //Return logP of s, and also set s.plog
            Int i,ilo,ihi,n1,n2;
            Doub st1,st2,stl1,stl2, ans;
            ilo = 0;
            ihi = ndat-1;
            while (ihi-ilo>1) { // Bisection to find where is t_c in the data
                            .
                            .
                            .
            ans =  n1*(s.k1*log(s.lam1)-factln(s.k1-1))+(s.k1-1)*stl1-s.lam1*st1;
//equation for log(p)
            ans += n2*(s.k2*log(s.lam2)-factln(s.k2-1))+(s.k2-1)*stl2-s.lam2*st2;
//equation for log(P)
            return (s.plog = ans);
      }
};
```

The Plog object is the only place that the data enter, and they enter only through the constructor. All other parts of the calculation see the data only through the calculation of $\log P$.

Part3

Next we come to the proposal generator. Almost any generator will work, but the timing is of the essence. The mean rate of recorded counts is λ/k.

Take into account that λ is continuous and at each step we will do a small changes of it, while k is discrete and every change is significant, especially at small k → if we naively write a generator that proposes independent changes in λ and k → the acceptable step in λ required for a change in k is so large that it is not probable → after we have settled down to roughly the right value of λ/k, essentially all proposals for changing k will be rejected. If we are not smart enough to recognize this problem ahead of time, we can find it experimentally by inspecting the Markov chain as it evolves and noting the proposals to change k are never accepted.

A solution is to have two kinds of steps:

1. Changes λ by a small amount, k is fixed,
2. Changes λ and k so that λ/k is fixed.

We choose randomly (but the first one is more probable) between the two kinds of steps.

The Proposal must return ratio $\frac{q(x_1|x_{2c})}{q(x_{2c}|x_1)}$ needed in $\alpha(x_1, x_{2c}) = \min\left(1, \frac{\pi(x_{2c})q(x_1|x_{2c})}{\pi(x_1)q(x_{2c}|x_1)}\right)$. Here is an example that proposes small lognormal steps for $\lambda_1, \lambda_2, t_c$, or else proposes incrementing k_1, k_2 by 1,0, or -1 with corresponding changes in λ for the second kind of step.

mcmc.h

```
struct Proposal { //Functor implementing the proposal distribution
        Normaldev gau;
        Doub logstep;

        Proposal(Int ranseed, Doub lstep) : gau(0.,1.,ranseed), logstep(lstep) {}

        void operator() (const State &s1, State &s2, Doub &qratio) { //Given state s1, set
state s2 to a proposed candidate. Also set qratio to q(s1|s2)/q(s2|s1)
                Doub r=gau.doub();
                if (r < 0.9) { //Lognormal steps holding the k's constant
                        s2.lam1 = s1.lam1 * exp(logstep*gau.dev());
                        s2.lam2 = s1.lam2 * exp(logstep*gau.dev());
                        s2.tc = s1.tc * exp(logstep*gau.dev());
                        s2.k1 = s1.k1;
                        s2.k2 = s1.k2;
                        qratio = (s2.lam1/s1.lam1)*(s2.lam2/s1.lam2)*(s2.tc/s1.tc);
//Factors for lognormal steps
                } else { //Steps that change k1 and/or k2
                        r=gau.doub();
                        if (s1.k1>1) {
                                if (r<0.5) s2.k1 = s1.k1;
                                else if (r<0.75) s2.k1 = s1.k1 + 1;
                                else s2.k1 = s1.k1 - 1;
                        } else { //k1=1 requires special treatment
                                if (r<0.75) s2.k1 = s1.k1;
                                else s2.k1 = s1.k1 + 1;
                        }
                        s2.lam1 = s2.k1*s1.lam1/s1.k1;
                        r=gau.doub(); //Now all the same for k2
                        if (s1.k2>1) {
                                if (r<0.5) s2.k2 = s1.k2;
                                else if (r<0.75) s2.k2 = s1.k2 + 1;
                                else s2.k2 = s1.k2 - 1;
                        } else {
                                if (r<0.75) s2.k2 = s1.k2;
                                else s2.k2 = s1.k2 + 1;
                        }
                        s2.lam2 = s2.k2*s1.lam2/s1.k2;
                        s2.tc = s1.tc;
                        qratio = 1.;
                }
        }
};
```

How shell we set `logstep`, the size of the proposed lognormal step? A rule of thumb for proposals like this with an adjustable scale is that the average acceptance probability ought to be in interval [.1, .4]. If it is much smaller, then decrease the step size parameter; if much larger increase the step size parameter. In our example, the value `logstep=.01` (proposed changes on the order of $\pm1\%$) gives good results.

Part4 – The only universal part of MCMC in this code

`mcmcstep` is a function that takes a specified number of steps implementing α. It has no persistent state and gets all the information it needs via the `State`, `Plog` and `Proposal` structures.

`mcmc.h`

```
Doub mcmcstep(Int m, State &s, Plog &plog, Proposal &propose) {//m - # of steps, choosing
vector x, plog and proposal functions. Take m MCMC steps, starting with and updating s
        State sprop; //Storage for candidate
        Doub qratio,alph,ran;
        Int accept=0;
        plog(s);
        for (Int i=0;i<m;i++) { //Loop over steps
                propose(s,sprop,qratio);
                alph = min(1.,qratio*exp(plog(sprop)-s.plog)); //alpha
                ran = propose.gau.doub();
                if (ran < alph) { //Accept the candidate
                        s = sprop;
                        plog(s);
                        accept++;
                }
        }
        return accept/Doub(m);
}
```

Part5 – main()

Assume $N = 1000$ data points t_i and start $x = \{\lambda_1 = 1, \lambda_2 = 3, t_c = 100, k_1 = 1, k_2 = 1\}$. Secretly, we know that the data were generated using the actual values $x = \{3,2,200,1,2\}$. The random seed is 10102, and the lognormal stepsize is .01. We'll take 1000 steps of burn-in and thereafter store values after every 10 steps.

We will also compute ergodic average quantities using:

- $\langle \lambda_1 \rangle = \frac{1}{n-k} \sum_{i=k}^{n-1} (\lambda_1)_i$,
- $k = 1000 - \#$ of burn-in steps,
- $n = k + 10000$ - # of steps that are averaged,
- $(\lambda_1)_i -$ value of λ_1 at i^{th} step.

`MCMC1.cpp`

```
#include "stdafx.h"
#include <iostream>
#include <algorithm>
#include <vector>
#include "nr3.h"
#include "ran.h"
#include "gamma.h"
#include "incgammabeta.h"
#include "deviates.h"
#include "mcmc.h"
using namespace std;

int main()
{
```

```
Doub avglam1, sumlam1 = 0, avglam2, sumlam2 = 0;
vector<Doub> arrlam1, arrlam2; //use vectors, declaring large arrays does not work

//Code from NR
Doub accept;
VecDoub times(1000);
// Fill the vector times here {t_0,...,t_(N-1)}
Int k1 = 1, k2 = 2, N = 1000; Doub lambda1 = 3.0, lambda2 = 2.0, tc = 200.0; //
Times are modeled by secretly knowing that the data were generated using the actual
values (3,2,200,1,2)
//If a Poisson process has a rate lambda, then the waiting time to the kth event
is distributed as Gamma(k,lambda)
Gammadev gdev1(k1, lambda1, 17); //17 is a randomly chosen seed
Gammadev gdev2(k2, lambda2, 17);
Doub t = 0.0;
for (Int i = 0; i<N; i++) {
        if (t < tc) { //before t_c
                t += gdev1.dev();
        }
        else { //after t_c
                t += gdev2.dev();
        }
        times[i] = t; //store all times in array to be used in plog
}
//End of vector times
State s(1., 3., 100., 1, 1); //s is x and we set initial parameters (lambda1,
lambda2, t_c, k_1, k_2)
Plog plog(times); // create object that calculates plog given vector times
Proposal propose(10102, .01); //10102 is a random seed, .01 is lognormal stepsize
for (int i = 0; i<1000; i++) accept = mcmcstep(1, s, plog, propose); //Burn - in,
we take 1000 steps of burn in
for (int i = 0; i<101000; i++) { // Production. Added 1000 because of the 1000
burn-in

        accept = mcmcstep(10, s, plog, propose); //store values after every 10
steps

        // Save values, increment averages, etc., here.
        //store values to vectors
        arrlam1.push_back(s.lam1);
        arrlam2.push_back(s.lam2);
}
for (int i = 1000; i < 101000; i++){
        sumlam1 = sumlam1 + arrlam1[i];
        sumlam2 = sumlam2 + arrlam2[i];
}
avglam1 = .00001*sumlam1;
avglam2 = .00001*sumlam2;
cout << "lambda1 average: " << avglam1 << endl;
cout << "lambda2 average: " << avglam2 << endl;
//observe how k1 and k2 have changed their values toward the end
cout << "k1: " << k1 << endl;
cout << "k2: " << k2 << endl;
cout << "t_c: " << tc;
return 0;
}
```

Output

```
lambda1 average: 3.01664
lambda2 average: 2.04496
k1: 1
k2: 2
t_c: 200
```

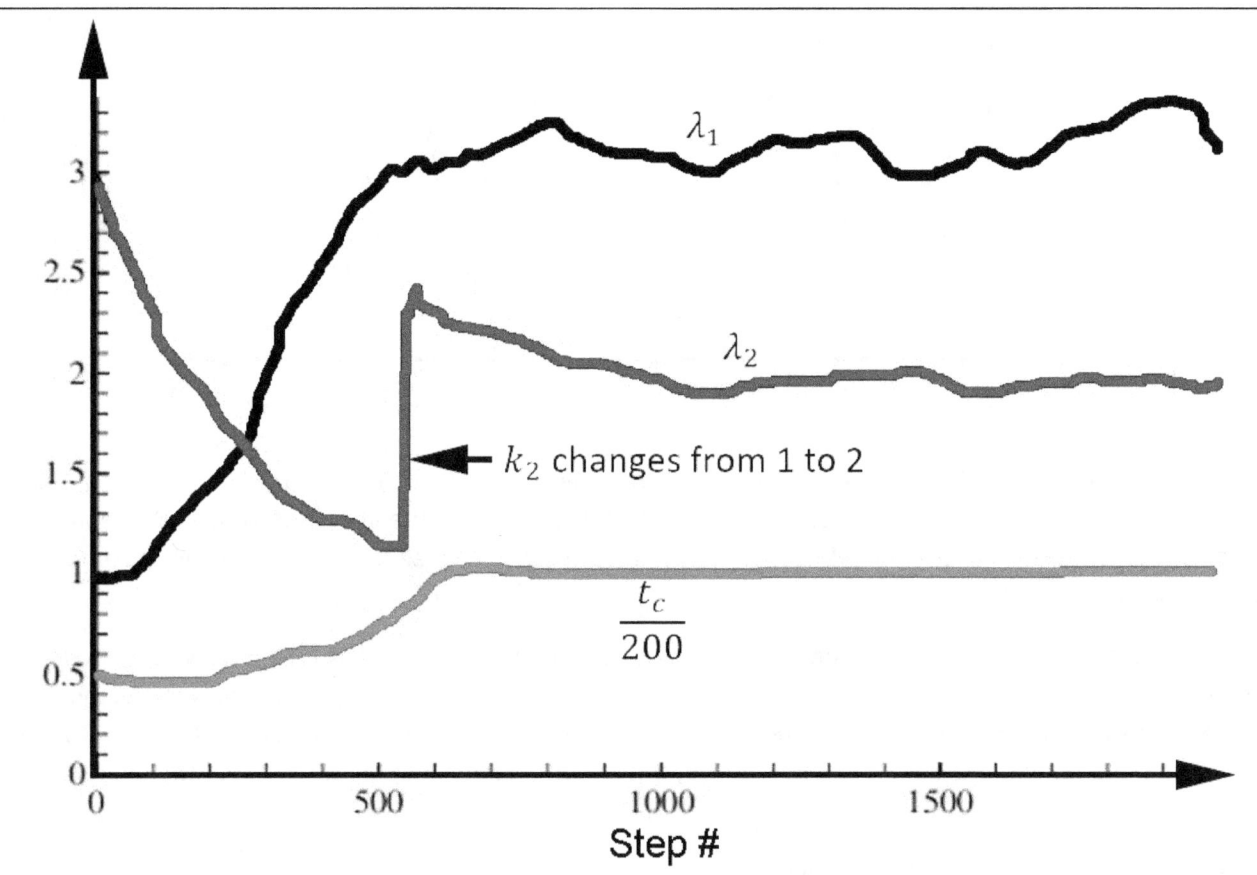

The figure shows the evolution of $\lambda_1, \lambda_2, t_c$ as a function of MCMC step. The burn-in time is seen to be ~1000 steps, after which the Markov chain explores the equilibrium distribution. During burn-in we can see the parameters heading toward equilibrium, with exception of λ_2, which goes rapidly to the value 1, with value $k_2 = 1$. These values replicate the mean rate of the recorded data. Only when it is near convergence does the model discover that the t_i's greater than t_c don't actually fit an exponential distribution (remember that exponential distribution is a special case of gamma distribution), but do fit a gamma distribution with the same mean rate but with $k_2 = 2$ (the correct answer). Had we not provided Proposal with a step that tests for this, we would have converged to a wrong answer. More precisely, we would have produced a model whose true burn-in time was very large.

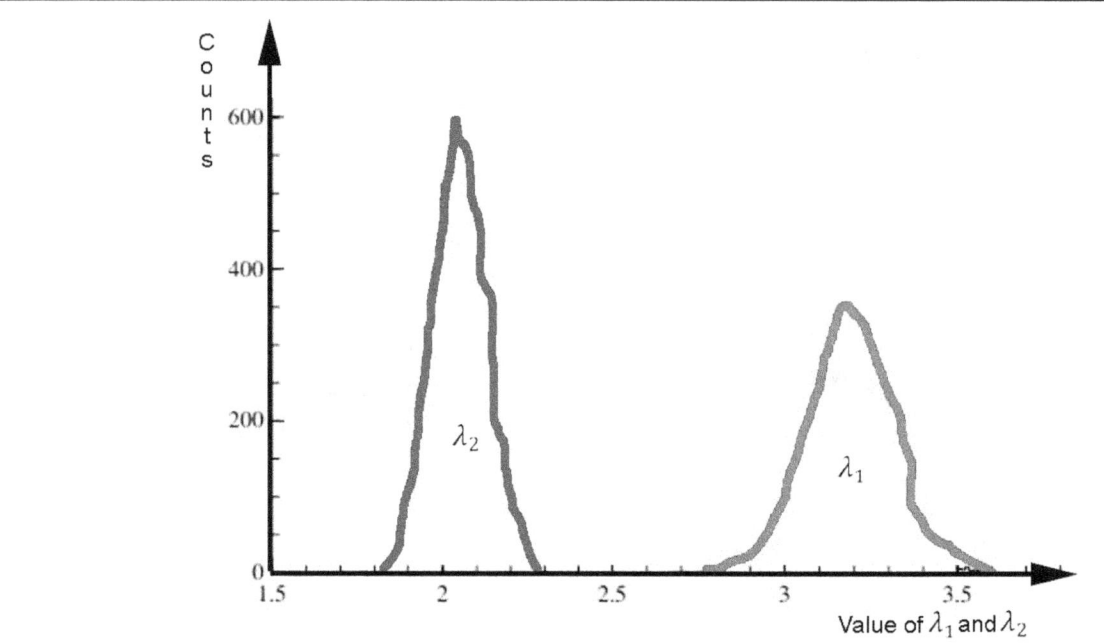

The figure shows the 10000 steps after burn-in, with values saved every 10th step, giving the histograms which represent the parameter values and their uncertainties. This is the payoff of MCMC: We learn not just about most likely parameter values, but also details about how well the parameters are determined by this particular data set.

By running the model for a long time, we could achieve precise distributions that have well-converged means. But they would not be centered on the true values 3 and 2. This is because the $\langle \lambda_1 \rangle$ is just the best apparent sample value of λ_1 for the particular data set. The relation of apparent value to the actual value has nothing to do with the standard error of $\langle \lambda_1 \rangle$, but is instead indicated by the width of the distribution of all the $(\lambda_1)_i$'s.

We should thus run an MCMC model only long enough to:

i)   be sure that there was a sufficient burn-in,

ii)   characterize distributions well enough that the error in mean quantities of interest is reasonably small compared with the observed dispersion of those quantities in the Markov chain.

9. Quantum Monte Carlo

QMC is a family of computational methods with aim the study of complex quantum systems.

Any physical system can be described by the many-body non-relativistic SE. The difficulty is that solving the SE requires the knowledge of the many-body wave function on the many-body Hilbert space, even so, it can be solved analytically only in a few highly idealized cases; for most realistic systems it must be solved numerically.

The properties of solids and molecules are determined almost exclusively by their electronic structure, described by the time-independent SE (TISE). One way to reduce computation time is to use Born-Oppenheimer approximation[12].

QMC provides an exact solution to the many-body problem for non-frustrated[13] interacting boson systems, and very accurate solution of interacting fermion systems.

The methods:[14]

- Variational MC (VMC), the simplest method, zero-temperature only,
- Diffusion MC (DMC), zero-temperature only,
- Green's function MC (GFMC),
- Path-Integral MC (PIMC), can be used for finite temperature,
- Auxiliary-field MC.

The VMC and DMC methods are best suited to calculating energies because these have the very advantageous zero-variance property; as the trial wave function approaches the energy eigenstate, the statistical fluctuations tend to zero. They have less well adapted to studying excited states but have nevertheless been used to do so.

PIMC is the main QMC method used to study bosonic systems such as liquid He. It is capable of simulating interacting electrons at finite temperature.

Statistical foundations

Here we will review the MC.

The validity of using MC for quantum calculations:

- M – total # of mesh points,
- N – total # of particles,
- $R = (r_1, \dots, r_N)$ – a walker, configuration, psip,
- $\{R_m, m = M\}$ – a set of uncorrelated configurations distributed according to $P(R)$

---

[12] Nuclei do not move.

[13] In condensed matter physics, the term frustration refers to a phenomenon, where atoms tend to stick to non-trivial positions or where, on a regular crystal lattice, conflicting inter-atomic forces (each one favoring rather simple, but different structures) lead to quite complex structures. As a consequence of the frustration in the geometry or in the forces, a plenitude of distinct ground states may result at zero temperature, and usual thermal ordering may be suppressed at higher temperatures.

[14] We will investigate only VMC and DMC.

- r_i – position of the ith particle,
- $P(R)$ – PDF of finding the particles in the configuration R, $\int P(R)dR = 1$ - normalized PDF,

- Quadrature methods → accuracy depends on the fineness of the integration mesh: d-dimensional cubic mesh used for d-dimensional integral → error scales as $M^{-\frac{4}{d}}$ → error decreases slowly for high dimensions.
- MC → regardless of dimensionality, statistical error decreases as the square rot of the number of sampling points (a consequence of the CLT).

Δ. We define a new random variable: $Z_f = \frac{\sum_{m=1}^{M} f(R_m)}{M}$

$f(R)$ has mean $\mu_f = \int f(R)P(R)dR$ and variance $\sigma_f^2 = \int \big(f(R) - \mu_f\big)^2 P(R)dR$ → the CLT tells us that for large M, Z_f is normally distributed with mean μ_f and standard deviation $\frac{\sigma_f}{\sqrt{M}}$ → regardless of $P(R)$ the mean value of a large # of measurements of some function of R will be a good estimator of the mean of that function with respect to $P(R)$, and the standard deviation will decrease as $\frac{1}{\sqrt{M}}$ irrespective of the dimension of the integral. ∎

This can be applied to the evaluation of integrals:

$$I = \int g(R)dR = \left| f = \frac{g}{P} \right| = \int fPdR = \lim_{M \to \infty} \left(\frac{1}{M} \sum_{m=1}^{M} f(R_m) \right) \approx \frac{1}{M} \sum_{m=1}^{M} f(R_m)$$

An estimate of the size of the error bar on the computed value of I is $\pm \frac{\sigma_f}{\sqrt{M}}$.

9.0 Trial wave functions

The trial wave function must well approximate an exact eigenstate for all R because it will improve calculation by:

- reducing the statistical uncertainty,
- lowering the energy toward the exact value,
- reducing the extent of any systematic error such as time-step error.

It should be:

- simple,
- compact,
- has simple easily evaluated 1st and 2nd derivatives,
- accurate everywhere.

The typical ψ_T of analytic variational calculations are not often useful, since they are severely restricted in form by the requirement that they be amenable to analytic integrations. QMC functions are not restricted in form since we integrate numerically, and even inter-electron distances r_{ij} can be included.

The typical trial wave function for QMC calculations consists of the product of a Slater determinant multiplied by a second function which accounts to some extent for electron correlation with use of inter-electron distances.

Fermionic trial wave function takes a form called Slater-Jastrow wave function:

$$\Psi_T = \Psi_D \Psi_J, \Psi_J = e^J$$

- Ψ_T – trial wave function,
- Ψ_D – Slater determinant,
- Ψ_J – Jastrow wave function,
- J – Jastrow factor.

9.0.1 Slater Determinant

Slater determinant is an expression that describes the wave function of multi-fermionic system that satisfies the anti-symmetry requirements and consequently the Pauli principle by changing sign upon exchange of two fermions.

- $x = (R, s_z)$ – position and spin,
- $\Psi(x)$ – spin-orbital wave function..

Two-particle system:

- Bosons: $\Psi(x_1, x_2) = \Psi(x_2, x_1) = \Psi(x_1)\Psi(x_2)$
- Fermions: $\Psi(x_1, x_2) = -\Psi(x_2, x_1) = \frac{1}{\sqrt{2}}\left(\Psi_1(x_1)\Psi_2(x_2) - \Psi_1(x_2)\Psi_2(x_1)\right)$

Many-particle system:

- Bosons (Hartree product): $\Psi(x_1, \dots, x_N) = \Psi_1(x_1) \cdots \Psi_N(x_N)$

- Fermions (Slater determinant): $\Psi(x_1, \ldots, x_N) = \begin{vmatrix} \Psi_1(x_1) & \cdots & \Psi_N(x_1) \\ \vdots & \ddots & \vdots \\ \Psi_1(x_N) & \cdots & \Psi_N(x_N) \end{vmatrix} = |\Psi_1 \cdots \Psi_N|$

For electron $\left(s = \frac{1}{2}\right)$ we can have spin-up or spin-down state, thus we can decompose Slater wave function (and remember that it is: $\Psi = \Psi_D \Psi_J$):

$$\Psi_D = \begin{vmatrix} \Psi_D^\uparrow(R^\uparrow) & 0 \\ 0 & \Psi_D^\downarrow(R^\downarrow) \end{vmatrix} = \Psi_D^\uparrow(R^\uparrow)\Psi_D^\downarrow(R^\downarrow)^{15}$$

- $\Psi_D^\uparrow = |\Psi_1 \cdots \Psi_{N^\uparrow}|$ - Slater determinant for electrons in spin-up,
- $\Psi_D^\downarrow = |\Psi_1 \cdots \Psi_{N^\downarrow}|$ - for electrons in spin-down,
- $\Psi_i(R)$ – orbitals,
- $N^\uparrow$ - # of electrons in spin-up,
- $N^\downarrow$ - in spin-down,
- $R^\uparrow = \{r_1, \ldots, r_{N^\uparrow}\}$ - coordinates for spin-up,
- $R^\downarrow = \{r_{N^\uparrow+1}, \ldots, r_N\}$ - for spin-down.

Example: Methane has 10 electrons $\rightarrow [\Psi_D]_{10\times10}$, and let's assume that it is $N^{\uparrow\downarrow} = 5$: $\left[\Psi_D^{\uparrow\downarrow}\right]_{5\times5}$.

The use of spin-independent wave functions is computationally advantageous because a large determinant is replaced by two smaller ones and no sums over spin variables are required.

The use of separate determinants for differing spins results in a trial function that is not antisymmetric with respect to interchange of opposite spin electrons, but gives the correct expectation values for spin-independent operators – the anti-symmetry of the wave function creates an exchange hole that keeps parallel-spin electrons apart, but there are no correlations between the positions of electrons with antiparallel spins. There is a significant probability of finding two antiparallel-spin electrons very close to each other $\rightarrow$ Coulomb repulsion energy is high. This problem can be solved by Jastrow wave function so that the overall wave function is antisymmetric.

Determinants are usually obtained by LDA-DFT or HF.[16]

The orbitals $\Psi_i(R)$, are expanded in computationally convenient basis-set, optimal for QMC with criterions speed of evaluation of the value and first and second derivatives at arbitrary points. The basis-set must be convenient for solving the single-particle equations. The orbitals must be real functions.

Typically we use:

- Radial numerical grids – for atomic systems,
- Gaussian function – for molecular systems,
- Plane waves – for continuum systems.

---

[15] Note that in some literature there is different and confusing notation: $\Psi_D = \det^{up}\det^{down}$
[16] LDA – Local-density approximation, DFT – Density functional theory, HF – Hartree-Fock.

9.0.2 Jastrow factor

The Jastrow factor is a function of all electron and atom positions.

A Jastrow factor should allow rapid evaluation, as this is one of the more computationally demanding parts of VMC and DMC calculations. In VMC it is the only method, in DMC it is not required, but is recommended since HF wave function is usually insufficiently accurate.

A good quality Jastrow factor should incorporate the physics of electron correlation in a compact and rapidly computable form that is convenient for optimization. Here we have a waging between the accuracy and speed. For example, a wave function obtaining 80% of the correlation energy and 30 parameters may be preferable to one obtaining 90% of the correlation energy with 200 parameters.

For DMC we prefer simpler forms. For VMC the preference depends on the specific application and desired accuracy.

First condition for wave function is that it is antisymmetric. Additional local set of constrains which may be readily imposed are for electron-electron and electron-nucleus interactions – these constrains are called "cusp conditions", a constraint on the derivatives of the wave function. For particle-particle coalescence:

$$\frac{d\Psi}{dr}\bigg|_{r=0} = \zeta\Psi|_{r=0}$$

- $r \equiv r_{ij} \lor r_{il}$[17]
- $\zeta = \begin{cases} -Z, \text{ coalescence of electron with a nucleus of charge } Z \\ \frac{1}{2}, \text{ electron-electron with parallel spins} \\ \frac{1}{4}, \text{ electron-electron with antiparallel spins} \end{cases}$

Most simulations retain only one- and two-body terms in J:

$$J(R) = \sum_{i=1}^{N} \chi(r_i) - \frac{1}{2}\sum_{i=1}^{N}\sum_{\substack{j=1 \\ j\neq i}}^{N} u(r_i, r_j)$$

- u – electron-electron correlations, with purpose to reduce the magnitude of the many-electron wave function whenever two electrons approach each other → reduces the probability of finding two electrons close together and decreases the electron-electron interaction energy → u was chosen to minimize the energy of the system: at places where interaction energy is low → u is high.
- χ – depends on the positions of the nuclei and describes the electron-nuclear correlation.

^4He

$$\psi = J = e^{\sum_{i<j}^{N} -u(r_{ij})}$$

---

[17] We label electrons with small letters, and nuclei with capital letters.

Solids – homogeneous electron gas

A commonly used and simple form suitable for solids:

$$J = e^{\sum_{i<j}^{N} -u_{\sigma_i,\sigma_j}(r_{ij})}, u_{\sigma_i,\sigma_j}(r_{ij}) = \frac{A_{\sigma_i,\sigma_j}}{r_{ij}}\left(1 - e^{-\frac{r_{ij}}{F_{\sigma_i,\sigma_j}}}\right)$$

The correlations between pairs of electrons depend on whether they have parallel or antiparallel spins → constants A_{σ_i,σ_j} and F_{σ_i,σ_j} are spin dependent.

Assuming that the solid is not spin polarized, there are 4 parameters:

    i. $A_{\uparrow\uparrow} = A_{\downarrow\downarrow}$,

    ii. $A_{\uparrow\downarrow} = A_{\downarrow\uparrow}$,

    iii. $F_{\uparrow\uparrow} = F_{\downarrow\downarrow}$,

    iv. $F_{\uparrow\downarrow} = F_{\downarrow\uparrow}$.

Two of these can be eliminated by imposing the cusp conditions (we replace F):

$$\left.\frac{du}{dr}\right|_{r=0} = \begin{cases} -\dfrac{1}{4}, \sigma_i = \sigma_j - \text{parallel spins} \\ \dfrac{1}{2}, \sigma_i \neq \sigma_j - \text{antiparallel spins} \end{cases} \Rightarrow \begin{cases} \dfrac{A_{\uparrow\uparrow}}{2F_{\uparrow\uparrow}^2} = \dfrac{1}{4} \\ \dfrac{A_{\uparrow\downarrow}}{2F_{\uparrow\downarrow}^2} = \dfrac{1}{2} \end{cases}$$

The remaining two parameters can be chosen to reproduce the long-range behavior:

$$\lim_{r_{ij}\to\infty} u = \frac{1}{\omega_p r_{ij}}, \omega_p = \sqrt{4\pi n} - \text{plasma frequency}, n - \text{electron number density} \Rightarrow A_{\uparrow\uparrow} = A_{\uparrow\downarrow} = \frac{1}{\omega_p}$$

Williamson's approach

Successful for: solid-state periodic boundary condition applications, like bulk carbon or silicon. It is an improvement on the solid state model given above. We model extended systems using small simulation cells subject to periodic boundary conditions. This periodicity must be reflected in u, which should remain unchanged if translated by a cell lattice vector.

Williamson has used u functions that tend to zero smoothly as r_{ij} approaches the radius r_{WS} of the largest sphere that fits inside the Wigner-Seitz[18] simulation cell → computationally efficient because the need for sums over images is eliminated:

$$u = 0, r_{ij} > r_{WS}$$

$$J = \sum_{i=1}^{N} \chi(r_i) - \sum_{i=1}^{N}\sum_{j\neq i}^{N} u(r_{ij})$$

- The one-body term consists of a plane-wave expansion: $\chi(r_i) = \sum_G \chi(G)e^{iGr}$, G – reciprocal lattice vectors.

---

[18] In a lattice (solid-state), volume-cell that is periodic and fills entire space.

- Electron-electron dependent term: $u_0(r) = \frac{A}{r}\left(1 - e^{-\frac{r}{F}}\right)e^{-\frac{r^2}{L_0^2}}$
  - $L_0 \approx 30\% \cdot r_{WS}$,
  - A – variational parameter,
  - F – chosen so that the cusp condition is obeyed.

Mitash' approach

Successful for DMC for: atomic systems, molecular carbon and silicon systems.

In real materials u no longer depends only on r_{ij}, but also on r_i and r_j. In atomic and molecular calculations it has become common to take account of this inhomogeneity by including three-body electron-electron-nucleus correlation terms:

$$J = \sum_I^N \sum_i^N \sum_{j \neq i}^N u(r_{iI}, r_{jI}, r_{ij})$$

- $I - I^{th}$ ion,
- $i - i^{th}$ electron of N electrons.

$$u(r_{iI}, r_{jI}, r_{ij}) = -\frac{c}{\gamma}e^{-\gamma r_{ij}} + \sum_{k<l.m} c_{klm}\left(a_k(r_{iI})a_l(r_{jI}) + a_k(r_{jI})a_l(r_{iI})\right)b_m(r_{ij})$$

- $a_k(r) = \left(\frac{\alpha_k r}{1+\alpha_k r}\right)^2, \alpha_k = \frac{\alpha_0}{2^{k-1}}, a_0 = 1$
- $b_m(r) = \left(\frac{\beta_m r}{1+\beta_m r}\right)^2, \beta_m = \frac{\beta_0}{2^{m-1}}, b_0 = 1$

9.0.3 Spin

It appears that the spin dependence of $\psi_T(x)$ complicates QMC algorithms, but this is not the case. Suppose we wish to use $\Psi(X)$ to calculate the expectation value of the spin-independent operator:

$$\langle O \rangle = \frac{\sum_\sigma \int \Psi^*(X)O(R)\Psi(X)dR}{\sum_\sigma \int \Psi^*(X)\Psi(X)dR}$$

For each spin configuration $\sigma = (\sigma_1, \ldots, \sigma_N)$, we can permute the arguments of $\Psi(X)$ (i.e. rearrange labels) such that the first $N_\uparrow$ have up-spin, and $N_\downarrow = N - N_\uparrow$ have down-spin:

$$\Psi(X) = \Psi(R, \sigma) = \Psi(\{r_1, \uparrow\}, \ldots, \{r_{N_\uparrow}, \uparrow\}, \{r_{N_\uparrow+1}, \downarrow\}, \ldots, \{r_N, \downarrow\}) \equiv \Psi(R')$$

O is not affected by the relabeling because it is symmetric with respect to exchange of electrons → the sums over spin configurations can be removed:

$$\langle O \rangle = \frac{\int \Psi^*(R')O(R')\Psi(R')dR'}{\int \Psi^*(R')\Psi(R')dR'}$$

9.1 VMC

VMC is a quantum MC method that applies the variational method to approximate the ground state of a quantum system. The main drawback is that the accuracy of the result depends entirely on the accuracy of the trial wave function.

VMC is the same as the conventional analytic variational method except that the required integrals are evaluated using MC. The procedure involves evaluation of the energy for a set of representative points in configuration space. The optimal values of the parameters is found upon minimizing the total energy.

The variance of the local energy of the wave function is a measure of its accuracy.

VMC has two types of errors:

1. Systematic error: due to the use of an approximate wave function,
2. Statistical uncertainty: due to the finite N.

9.1.1 Review of the Variational method

Variational method is approximate method for finding the ground state energy and wave functions. The true art of the variational method is in finding the good trial wave function by physical insight.

Requirements:

   i. Ψ_T must be continuous wherever the potential is finite,
   ii. $\nabla\Psi_T$ must be continuous wherever the potential is finite,
   iii. $\int \Psi_T^* \Psi_T$ must exist,
   iv. $\int \Psi_T^* \mathcal{H} \Psi_T$ must exist,
   v. $\int \Psi_T^* \mathcal{H}^2 \Psi_T$ must exist.

Method:

1. Choose a trial wave function with several unknown parameters: $\{\alpha_1, \dots, \alpha_n\}$
2. Find the parameters for which the expectation value of the energy is the lowest possible:

$$\frac{\partial E}{\partial \alpha_1} = \cdots \frac{\partial E}{\partial \alpha_r} = 0$$

3. The wave function is then an approximation to the ground state wave function, and the expectation value of the energy is an upper bound to the ground state energy.

For a given Hamiltonian: $E = \dfrac{\langle \Psi | \mathcal{H} | \Psi \rangle}{\langle \Psi | \Psi \rangle}$

- $E \geq E_0$, where E_0 is the ground state,
- $E = E_0$ iff Ψ is equal to the wave function of the ground state.

Example1: Gaussian trial function

We are given:

- $V(x) = \alpha|x|$
- $\psi = Ae^{-bx^2}$

$$\langle\psi|\psi\rangle = |A|^2 \int_{-\infty}^{\infty} e^{-2bx^2} dx = 1 \Rightarrow A = \left(\frac{2b}{\pi}\right)^{\frac{1}{4}}$$

$$E = \sqrt{\frac{2b}{\pi}} \int_{-\infty}^{\infty} \left(-\frac{\hbar^2}{2m} e^{-bx^2} \frac{d^2}{dx^2}\left(e^{-bx^2}\right) + e^{-2bx^2} \alpha|x|\right) dx$$

$$= 2\sqrt{\frac{2b}{\pi}} \int_{0}^{\infty} \left(-\frac{\hbar^2}{2m} e^{-bx^2} \underbrace{\frac{d^2}{dx^2}\left(e^{-bx^2}\right)}_{(2bx^2-1)2be^{-bx^2}} + e^{-2bx^2} \alpha x\right) dx$$

$$= 2\sqrt{\frac{2b}{\pi}} \int_{0}^{\infty} \left(-\frac{\hbar^2 b}{m} e^{-2bx^2}(2bx^2 - 1) + e^{-2bx^2} \alpha x\right) dx = \frac{1}{2\sqrt{2\pi m}}\left(\hbar^2 b\sqrt{2\pi} + \frac{2\alpha m}{\sqrt{b}}\right)$$

$$\frac{\partial E}{\partial b} = \frac{1}{2\sqrt{2\pi m}}\left(\hbar^2\sqrt{2\pi} - \alpha m b^{-\frac{3}{2}}\right) = 0 \Rightarrow b = \frac{(\alpha m)^{\frac{2}{3}}}{(2\pi)^{\frac{1}{3}} \hbar^{\frac{4}{3}}} \Rightarrow E_0 \le \frac{3(2\alpha\hbar)^{\frac{2}{3}}}{4(\pi m)^{\frac{1}{3}}}$$

Example2: 2D LHO

This example is just to demonstrate the method. Let us use the test wave function: $\psi_\alpha = A\rho^2 e^{-\alpha\rho}$

$$\mathcal{H} = -\frac{\hbar^2}{2m}\left(\frac{1}{\rho}\frac{\partial}{\partial\rho}\left(\rho\frac{\partial}{\partial\rho}\right)\right) + \frac{1}{2}m\omega^2\rho^2$$

$$\langle\psi_\alpha|\psi_\alpha\rangle = \frac{15}{4}\frac{A^2\pi}{\alpha^6} = 1 \Rightarrow A = \sqrt{\frac{4\alpha^6}{15\pi}}$$

$$E(\alpha) = \frac{\langle\Psi_\alpha|\mathcal{H}|\Psi_\alpha\rangle}{\langle\Psi_\alpha|\Psi_\alpha\rangle} = 2\pi A^2 \int_{0}^{\infty} \rho d\rho \rho^2 e^{-\alpha\rho} \mathcal{H}\rho^2 e^{-\alpha\rho} d\rho = \frac{\hbar^2}{10m}\alpha^2 + \frac{m\omega^2}{4}\frac{21}{\alpha^2}$$

$$\frac{\partial E(\alpha)}{\partial\alpha} = 0 \Rightarrow \alpha^4 = \frac{105}{2}\frac{m^2\omega^2}{\hbar^2} \Rightarrow \alpha = \left(\frac{105}{2}\right)^{\frac{1}{4}}\sqrt{\frac{m\omega}{\hbar}}$$

$$E(\alpha) = \hbar\omega\sqrt{\frac{73}{10}}, \psi = \sqrt{\frac{4}{15\pi}}\left(\frac{105}{2}\right)^{\frac{3}{4}}\left(\frac{m\omega}{\hbar}\right)^{\frac{3}{2}}\rho^2 e^{-\left(\frac{105}{2}\right)^{\frac{1}{4}}\sqrt{\frac{m\omega}{\hbar}}\rho}$$

9.1.2 VMC using Simple MC

We will be using the example 1 from above, but instead of analytically integrating, we will be using Simple MC method.

Just to visualize the wave function, we will set $\alpha = 1$ and work in atomic units, i.e. $m = \hbar = 1$:

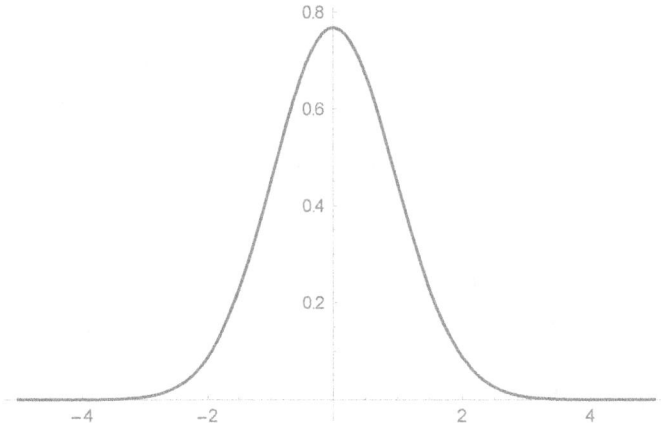

Almost all information lies within $(-4, 4)$.

Let us rework the previous example:

- $V(x) = |x|$
- $\psi = Ae^{-bx^2}$
- $A = \left(\dfrac{2b}{\pi}\right)^{\frac{1}{4}}$

Integrating in Wolfram Mathematica we get the exact result:

$$E = \frac{2}{3}\sqrt{\frac{2}{3}}b + \frac{1}{\sqrt{2\pi b}}$$

however, we want to use Simple MC to evaluate the integral:

$$E = 2\sqrt{\frac{2b}{\pi}} \int_0^\infty \left(-\frac{1}{2}e^{-bx^2}\frac{d^2}{dx^2}\left(e^{-bx^2}\right) + e^{-2bx^2}x\right)dx$$

$$= 2\sqrt{\frac{2b}{\pi}} \int_0^\infty \left(-be^{-2bx^2}(2bx^2 - 1) + e^{-2bx^2}x\right)dx$$

We need to get rid of the parameter b from the integrals in order to get a numerical solution. We can do this by substitution: $y = \sqrt{b}x$:

$$E = 2\sqrt{\frac{2b}{\pi}} \int_0^\infty \left(-be^{-2y^2}(2y^2 - 1) + \frac{e^{-2y^2}y}{\sqrt{b}} \right) \frac{dy}{\sqrt{b}}$$

$$= 2\sqrt{\frac{2}{\pi}} \left(-b \int_0^\infty e^{-2y^2}(2y^2 - 1)dy + \frac{1}{\sqrt{b}} \int_0^\infty e^{-2y^2}ydy \right)$$

Now we have 2 integrals which we can integrate using Simple MC:

1. $\int_0^\infty (2y^2 - 1)e^{-2y^2}dy$
2. $\int_0^\infty ye^{-2y^2}dy$

By plotting these two, we can see that almost all points lie in the interval (0,2).

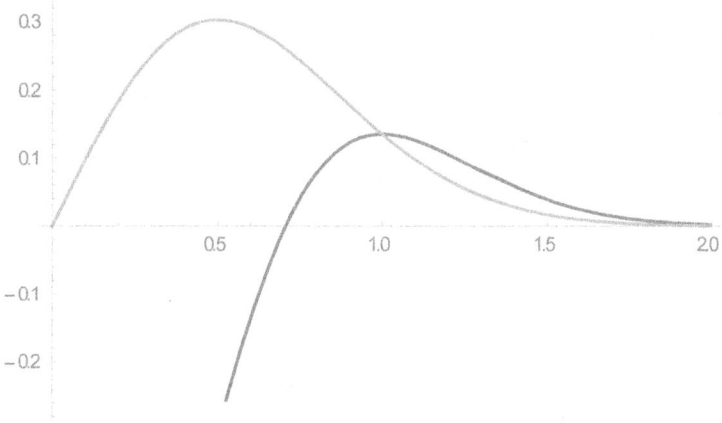

We will evaluate them one by one and we will use what we know in order to make a Simple MC efficient: the region of integration. If we would to sample from a larger interval vast majority of points would be wasted. We cheated here by using Wolfram Mathematica.

Code

In `mcintegrate.h` we will change the name from `torusfuncs` to `funcs` and from `torusregion` to `region`. That part will look like this for the first integral:

```
VecDoub funcs(const VecDoub &x) {
        Doub den = (2*pow(x[0],2)-1)*exp(-2*pow(x[0],2)); //Our integral will be the density
        VecDoub f(1);
        f[0] = den;
        return f;
}

Bool region(const VecDoub &x) {
        return x[0]<=2; //We will integrate in the most important interval (0,2)
}
```

For the second integral change the line 62 to:

```
Doub den = x[0] * exp(-2 * pow(x[0], 2));
```

VMC_Simple.cpp

```cpp
#include "stdafx.h"
#include <iostream>
#include "nr3.h"
#include "ran.h"
#include "mcintegrate.h"
using namespace std;

int main()
{
        VecDoub xlo(1), xhi(1); //give xlo and xhi to be in 1D
        xlo[0] = 0.; xhi[0] = 2.; // x coordinates

        MCintegrate mymc(xlo, xhi, funcs, region, NULL, 10201);

        mymc.step(1000000);
        mymc.calcanswers();
        // Display results
        cout << "\nmymc.ff = " << mymc.ff; // Display integral
        cout << "\nmymc.fferr = " << mymc.fferr; //Display error
        return 0;
}
```

Output for the integrals:

1. `mymc.ff = -0.313851, mymc.fferr = 0.000715049`

2. `mymc.ff = 0.249855, mymc.fferr = 0.000219797`

Using first output:

$$E = 2\sqrt{\frac{2}{\pi}}\left(-bI_1 + \frac{1}{\sqrt{b}}I_2\right)$$

$$\frac{\partial E}{\partial b} = 2\sqrt{\frac{2}{\pi}}\left(-I_1 - \frac{I_2}{2b^{\frac{3}{2}}}\right) = 0 \Rightarrow b = \frac{1}{2^{\frac{2}{3}}\left(-\frac{I_1}{I_2}\right)^{\frac{2}{3}}} \approx .541115$$

The exact solution is: $b = \frac{1}{(2\pi)^{\frac{1}{3}}} \approx 0.541926$, to which we got very close.

9.1.3 VMC using MCMC

- $R = (r_1, \dots, r_N)$ – many-body configuration,
- r_k – position of the k-th particle,
- a – unknown parameters,
- $E_L(R)$ – local energy, depending on the 3N coordinates,
- P – normalized PDF which mimics the density of more complex Ψ_0^2.

Sample points of unit weight are obtained with probabilities proportional to the P i.e. Metropolis algorithm is used to sample a set of points $\{R_m\} \sim P \rightarrow r_k$ are sampled using the Metropolis algorithm.

E_L has the useful property that for an exact eigenstate of the Hamiltonian: $E_L = $ const. For a general trial wave function: $E_L \neq$ const. and the variance of the E_L is a measure of how well the Ψ approximates an eigenstate.

$$\langle E \rangle = \frac{\langle \Psi(a)|\mathcal{H}|\Psi(a)\rangle}{\langle \Psi(a)|\Psi(a)\rangle} = \frac{\int |\Psi(R,a)|^2 \frac{\mathcal{H}\Psi(R,a)}{\Psi(R,a)} dR}{\int |\Psi(R,a)|^2 dR} = \int \underbrace{\frac{|\Psi(R,a)|^2}{\int |\Psi(R,a)|^2 dR}}_{P} \underbrace{\frac{\mathcal{H}\Psi(R,a)}{\Psi(R,a)}}_{E_L} dR = \int E_L P dR$$

If we remember that the expectation of random function is:

- discrete: $\langle f(x) \rangle = \sum f(x)P(x)$
- continuous: $\langle f(x) \rangle = \int f(x)P(x)dx$

We can see that $\langle E \rangle$ can be estimated as the average value of the local energies on a sample of N points R_k sampled from the $P(R)$:

$$\langle E \rangle \approx \langle E_L \rangle = \frac{1}{N} \sum_{k=0}^{N-1} E_L(r_k)$$

The advantage of this approach is that it does not use an analytical integration involving the wave function, and thus does not impose severe constraints on the form of the wave function.

Algorithm

In this algorithm, the electrons are moved individually and not as a whole configuration. This improves the efficiency of the algorithm in larger systems, where configuration moves require increasingly small steps to maintain the acceptance ratio.

VMC algorithm consists of two distinct phases:

1. (A walker consisting of an initially random set of electron positions) is propagated according to Metropolis algorithm, in order to equilibrate it and begin sampling $|\Psi|^2$.
2. The walker continues to be moved, but energies and other observables are also accumulated for later averaging and statistical analysis.

In more detail:

1. Equilibrium phase:
    a. Generate initial configuration using random positions for electrons.
    b. For every electron:
        i. Propose a move from r to r',
        ii. Compute $w = \left|\frac{\Psi(r')}{\Psi(r)}\right|^2$, the weight of the new-to-old,
        iii. Accept or reject the move according to Metropolis probability: $\min(w, 1)$. If the move is not accepted, the old point is treated as a new point.

     c.   Repeat configuration moves until equilibrated.

2. Accumulation phase:

     a.   For every electron in the configuration:

         i.   Propose a move from r to r',

        ii.   Compute $\left|\dfrac{\Psi(r')}{\Psi(r)}\right|^2$,

       iii.   Accumulate the contribution to the local energy, and other observables, at r' and at r, weighted by the Metropolis acceptance and rejection probabilities,

       iv.   Accept or reject move according to Metropolis acceptance probability.

     b.   Repeat configuration moves until sufficient data is accumulated.

Or with the flow-chart:

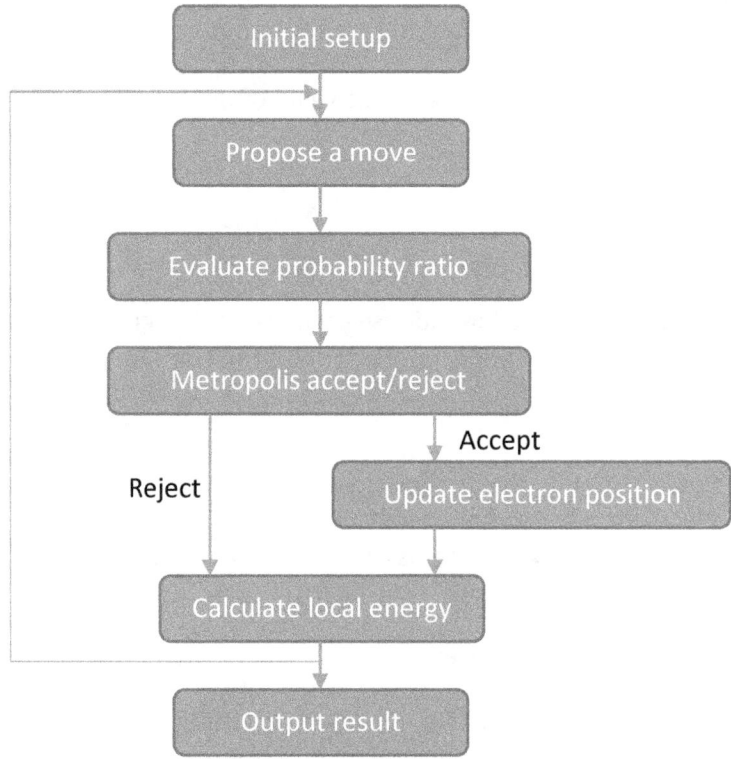

Observables are accumulated on a per electron basis, weighted by the acceptance and rejection probabilities:

$$\langle O \rangle = \frac{1}{N}\sum_{i=1}^{N}\left(p_i O_i(r') + (1 - p_i)O_i\right)$$

- O – observable,
- N - # of steps,
- p_i – Metropolis acceptance probability for i^{th} electron,
- $(1 - p_i)$ – rejection probability,
- O_i – contribution to the value of the observable.

9.2 DMC

DMC overcomes the limitation that VMC has (dependence on the trial wave function) by using a stochastic projection technique to enhance the ground-state component of a starting trial wave function.

In DMC the wave function of a system is represented by the density of many walkers diffusing throughout configuration space. A time evolution operator controls the motion, projecting out the ground state eigenfunction from all possible wave functions. The efficiency of the technique is improved by using a guiding wave function to herd walkers into important regions.

- $R = (r_1, \dots, r_N)$ – vector that describes the positions of N electrons, $\dim R = 3N$,
- r_k – position of the kth particle, named configuration, walker or psips.
- N - # of random samples of configuration space, forming
- $\{R_n\}_{n=1}^N$ – a sequence distributed according to
- $\rho(R)$ – a probability distribution.
- $\Psi(R,t)$ – time dependent wave function,
- $\psi(R)$ – time independent wave function.

The sequence is used to perform some mathematical or algorithmic operation:

$$F = \int f(R)\rho(R)dR \approx \frac{1}{N}\sum_{n=1}^{N} f(R_n)$$

SE: $\mathcal{H}\Psi = E\Psi, E = i\hbar\frac{\partial}{\partial t}, \mathcal{H} = -\frac{\hbar^2}{2m}\nabla^2 + V(R)$[19]

Switch to imaginary time by $\tau = it$: $\frac{\partial\Psi}{\partial\tau} = -\mathcal{H}\Psi$

Let us now switch to atomic units: $\frac{\partial\Psi}{\partial\tau} = \left(\frac{1}{2}\nabla^2 - V\right)\Psi$

$\Psi = \psi e^{-E\tau}$, where $e^{-E\tau}$ is the time evolution operator.

We can expand the state Ψ in eigenstates of the Hamiltonian E_i:

$$\Psi = \sum_{i=0}^{\infty} c_i\phi_i(R)e^{-E_i\tau}, \mathcal{H}\phi_i = E_i\phi_i$$

Ground-state: $\mathcal{H}\phi_0 = E_0\phi_0$

For any physical system, these eigenvalues must form an increasing set of real values, with E_0 the lowest member: $E_0 < E_1 \leq \cdots$

This means that the time evolution operator will exponentially **damp** all eigenfunctions ϕ_i, the damping being stronger for higher states → for large τ: $\lim_{\tau\to\infty} \Psi = c_0\phi_0 e^{-E_0\tau} = 0$ – this is the problem, which we can avoid by adding an arbitrary reference energy E_T and making the limit finite:

---

[19] Note that: $\nabla^2 \equiv \sum_{i=1}^{N} \nabla_i^2$

$$\frac{\partial \Psi}{\partial \tau} = \left(\frac{1}{2}\nabla^2 - V + E_T\right)\Psi \Rightarrow \lim_{\tau \to \infty} \Psi = c_0 \phi_0 e^{-(E_0 - E_T)\tau}$$

9.2.1 The diffusion analogue

Diffusion equation is isomorphic to the imaginary time SE:

$$\frac{\partial \rho(r,t)}{\partial t} = \left(D\nabla^2 + A(r)\right)\rho \Leftrightarrow \frac{\partial \Psi}{\partial \tau} = \left(\frac{1}{2}\nabla^2 - V + E_T\right)\Psi$$

- $\rho(r,t)$ - spatial density: $\rho \equiv \Psi$
- D – diffusion constant, particles moving a distance ΔR at positive or negative direction at interval $\Delta \tau$: $D = \frac{\overline{(\Delta R)^2}}{2\Delta \tau}$, or in our case: $D \equiv \frac{1}{2}$
- $A(r)$ – branching term, representing time-invariant sources/sinks: $A \equiv -(V - E_T)$

We would usually solve this diffusion equation and get particles diffusing under random motion. In DMC it is the other way around: we use random walkers diffusing through configuration space to describe a solution of the imaginary time SE.

Our configuration changes: $(R, \tau) \to (R', \tau + \Delta \tau) \to \Psi(R', \tau + \Delta \tau) = \int \Psi(R, \tau) G(R \to R', \Delta \tau) dR$

Replace Ψ by a set of delta functions, each representing one walker: $\Psi(R, \tau) = \sum_k \delta(R - R_k) \wedge \Psi(R', \tau + \Delta \tau) = \int \Psi(R, \tau) G(R \to R', \Delta \tau) dR \to \sum_k \delta(R' - R'_k) = \int \sum_k \delta(R - R_k) G(R \to R', \Delta \tau) dR = \sum_k G(R_k \to R'_k, \Delta \tau) \to \delta(R' - R'_k) = G(R_k \to R'_k, \Delta \tau) - G$ can be interpreted as the transition probability:

$$G(R \to R', \Delta \tau) = \left\langle \Psi(R', \tau + \Delta \tau) \middle| e^{-\tau(\mathcal{H} - E_T)} \middle| \Psi(R, \Delta \tau) \right\rangle$$

The Green's function may be approximated to second order in τ by factorizing the propagator into branching G_B and diffusion G_D parts:

$$\Delta \tau \ll : G(R \to R', \Delta \tau) = G_D(R \to R', \Delta \tau) G_B(R \to R', \Delta \tau) + O(\Delta \tau^3)$$

- $G_D(R \to R', \Delta \tau) = (2\pi\Delta\tau)^{-\frac{3N}{2}} e^{-\frac{(R'-R)^2}{2\Delta\tau}}$ – Gaussian function in R' with $\mu = R$, the transition probability for walkers with no external potential, i.e. only due to diffusion process.
- $G_B(R \to R', \Delta \tau) = e^{-\frac{\Delta\tau}{2}(V(R)+V(R')-2E_T)}$ – renormalization factor related to the branching process caused by external potential $V(R)$. In practice G_B is not normally included in the equation, but instead the branching process is taken into account by increasing or decreasing the population of walkers in the simulation. In its simplest form[20]:
  - $G_B < 1 \to$ the walker continues its evolution with probability G_B,
  - $G_B \geq 1 \to$ the walker continues, in addition, at the same position, a new walker is created with probability $(G_B - 1)$.

Outline of the DMC procedure:

1. distribute many walkers randomly: R,
2. advance by $\Delta \tau$ and move walkers to R' with a probability: $G_D(R \to R', \Delta \tau)$,

---

[20] More about this will be explained in the Algorithm.

3. keep going until for practical purposes all terms in $\Psi(R, \tau)$ other than the ground state have been damped out.

Examples

The following examples have wave functions that do not take into account the spin. Later, when fixed-node approximation is introduced, we will take it into account.

1D LHO

$$V = \frac{1}{2}kx^2, k = m\omega^2$$

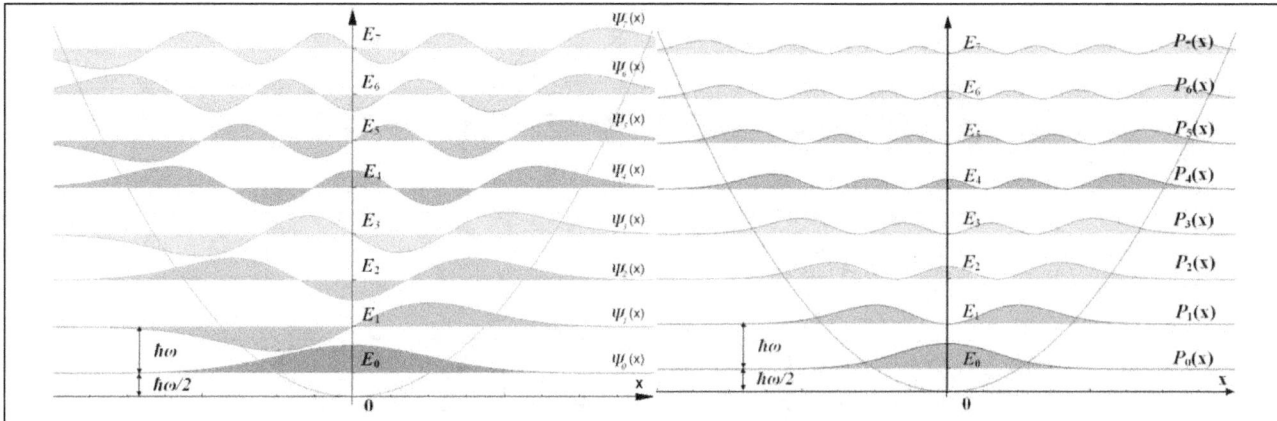

Left: wave functions for the eigenstates $n = 0 \dots 7$. Right: corresponding probability densities.

V may be shifted by an arbitrary constant energy E_T to make V negative in the central region, near $x = 0$: $V \to V - E_T$.

Initial collection of walkers is distributed about $x = 0$. We move each walker right or left by $(\Delta x, \Delta \tau)$, during which they diffuse, and give birth or die with probability (for each walker):

$$P_{b/d} = |V - E_T|\Delta\tau$$

In the negative region they multiply, at the positive they die. The changes for birth rise as we approach $x = 0$, and for the death as we move further from the center: the larger the potential energy, the more probable is the death; the lower the (negative) potential energy is, the higher is the chance for birth.

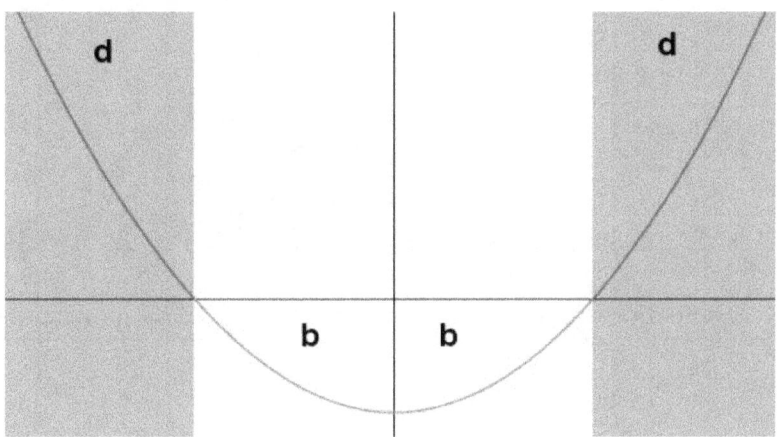

After large # of steps, distribution of walkers approaches a steady-state distribution:

$$\psi_0 \propto e^{-ax^2}, a = \frac{1}{2}\sqrt{k}$$

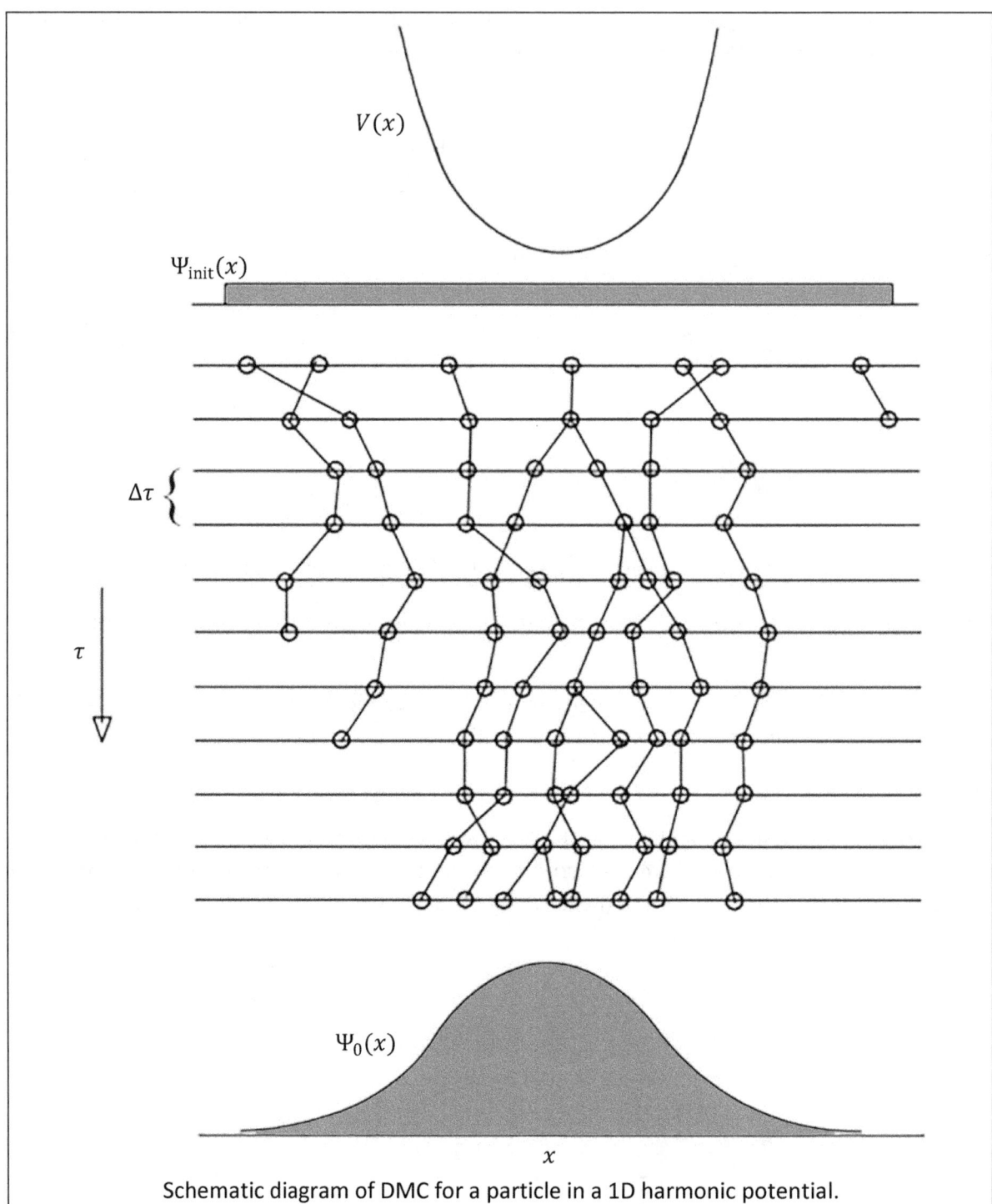

Schematic diagram of DMC for a particle in a 1D harmonic potential.

H_2

Labels for electrons: $\{1,2\}$, and for protons: $\{A, B\}$:

$$\frac{\partial \Psi}{\partial \tau} = \frac{1}{2}(\nabla_1^2 + \nabla_2^2)\Psi - V\Psi$$

$$V = -\frac{1}{r_{1A}} - \frac{1}{r_{1B}} - \frac{1}{r_{2A}} - \frac{1}{r_{2B}} + \frac{1}{r_{12}}$$

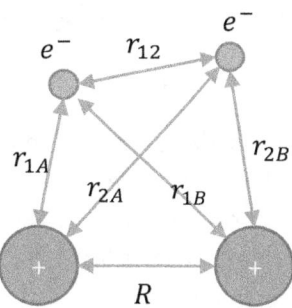

Shift the energy by the reference energy:

$$\frac{\partial \Psi}{\partial \tau} = \frac{1}{2}(\nabla_1^2 + \nabla_2^2)\Psi - (V - E_T)\Psi$$

A calculation is begun with > 1000 walkers in positions corresponding to electron configurations so that it is allowed to approach the steady-state distribution.

The configuration space is in 6D, so the random walk is in 6D space, by the non-uniform Δx in each dimension:

$$\Delta x \sim P(\Delta x) = \frac{1}{\sqrt{2\pi}\sigma} e^{-\frac{(\Delta x)^2}{2\sigma^2}} {}^{21}$$

After each move, for each walker:

$$P_{b/d} = |V - E_T|\Delta \tau \Rightarrow \begin{cases} u < P_b - \text{birth} \\ u > P_b - \text{nothing} \\ u < P_d - \text{death} \\ u > P_d - \text{nothing} \end{cases}, u \sim U(0,1)$$

In order to maintain the # of walkers approximately constant, E_T may be adjusted occasionally.

$$\frac{\partial \Psi}{\partial \tau} = -E\Psi \Rightarrow E = -\frac{1}{\Psi}\frac{\partial \Psi}{\partial \tau} \wedge N \propto \Psi \Rightarrow E = -\frac{1}{N}\frac{\partial N}{\partial \tau}$$

H_3^+

It has no boundaries serving as sinks or sources for walkers → the total # of walkers is not directly affected by the diffusion terms, but changes according to:

$$\frac{\partial N}{\partial \tau} = -\sum_N (V - E_T) \Rightarrow E = \langle V \rangle$$

---

[21] Note the similarities between $P(\Delta x)$ which was given for 1D, and G_D in 3D: $\sigma^2 \equiv \Delta \tau$.

9.2.2 Markov and Metropolis

Walkers are distributed with Metropolis algorithm according to: $\Psi(R, \tau) = \sum_k \delta(R - R_k)$.

Let's review Markov chain and Metropolis, but in the light of the DMC.

We wish to generate a sequence (chain) of configurations $\{R_m\}$, distributed according to some probability distribution $\rho(R)$. Techniques for generating such a sequence require $\rho(R)$ to be normalized – a problem if the wave function is unknown. The Metropolis constructs a Markov chain of configurations, with each configuration related to the last. The algorithm ensures that this chain is representative of $\rho(R)$, without the need for normalization.

Markov chains

- $P_i(R)$ – probability of the configuration R occurring, in a way #i.
- $T(R \rightarrow R')$ - transition probability,
- $\sum_{R'} T(R \rightarrow R') = 1$ – transition probabilities are normalized.

1. Uncorrelated chain with all members of the chain generated independently → the probability of obtaining a given sequence of N configurations R_1 to R_N:
$$P_N(R_1, \dots, R_N) = P_1(R_1) \dots P_1(R_N)$$

2. Markov chain: the transition is defined so that the next configuration is generated from the last, for example a random walk:
$$P_N(R_1, \dots, R_N) = P_1(R_1)T(R_1 \rightarrow R_2) \dots T(R_{N-1} \rightarrow R_N)$$

Ergodic Markov chain satisfies:

1. Every configuration in the sequence can be accessed from every other configuration in a finite number of steps.
2. There is no periodicity in the sequence.

For long enough times, the configurations are independent of the position within the sequence and the initial configuration.

Metropolis

Master equation which describes how the distribution changes:

$$\rho(R, \tau + \Delta\tau) - \rho(R, \tau) = -\underbrace{\sum_{R'} T(R \rightarrow R')\rho(R, \tau)}_{\substack{\rho(R) \text{ decreased by} \\ \text{configurations leaving} \\ R(\tau) \text{ for } R'(\tau + \Delta\tau)}} + \underbrace{\sum_{R'} T(R' \rightarrow R)\rho(R', \tau)}_{\substack{\rho(R) \text{ increased by} \\ \text{configurations joining} \\ R(\tau + \Delta\tau) \text{ from } R'(\tau)}}$$

When the distribution has become stationary:
$$\rho(R, \tau + \Delta\tau) - \rho(R, \tau) = 0 \Leftrightarrow$$
$$\rho(R, \tau + \Delta\tau) = \rho(R, \tau) \Leftrightarrow$$
$$\sum_{R'} T(R \rightarrow R')\rho(R, \tau) = \sum_{R'} T(R' \rightarrow R)\rho(R', \tau)$$

There are many solutions to this equation, but a particular solution is just:

$T(R \rightarrow R')\rho(R) = T(R' \rightarrow R)\rho(R'), (\forall R, R')$ - detailed balance equation – the rate at which walkers move between different regions of configuration space is the same, so that $\rho(R)$ remains stationary.

Analogy:

- $\rho(R)$ – amount of water in bucker R,
- $\rho(R')$ – amount of water in bucker R',
- $T(R \rightarrow R')\rho(R)$ – pumping rate from R to R'

$T(R \rightarrow R')\rho(R) = T(R' \rightarrow R)\rho(R'), (\forall R, R')$ - the pumping rate between both buckets is equal, so the amount of water in both buckets will remain the same.

The detailed balance condition is satisfied by the algorithm:

1. Start with a random configuration R.
2. Make a trial move to R', according to a transition probability $T(R \rightarrow R')$.
3. Accept the move with probability:

$$A(R \rightarrow R') = \min\left(\frac{T(R' \rightarrow R)\rho(R')}{T(R \rightarrow R')\rho(R)}, 1\right)$$

   a. trial move is accepted → R' is the next configuration in the walk,
   b. rejected → R is the next – the next configuration is identical to the current configuration. If $\rho(R)$ is high → $A(R \rightarrow R')$ will be low → R will occur often in the walk.
4. Return to step 2.

After an equilibration phase during which walkers settle down, $\{R_n\}$ will be distributed according to $\rho(R)$.

9.2.3 Importance sampling

DMC as described above is inefficient, since G_B can fluctuate massively between steps or even diverge, leading to massive fluctuations in walker populations. The first application in importance sampling MC was made for nodeless ground state of H_3^+, and the result was that the error was twenty times reduced. Without this improvement, DMC involving hundreds or thousands of electrons would not be possible.

In importance sampling DMC, guide functions ψ_T are used to encourage walkers into regions where the wave function is significant. It is possible to obtain very high accuracies by extending DMC calculations to calculate corrections to trial wave functions rather than the complete wave function.

For a good ψ_T the local energy is close to the ground-state energy eigenvalue and roughly constant, so the population fluctuations are much diminished.

$\psi_T = 0$ where $V \rightarrow \infty$ or at the positions of the nuclei.

Instead of $\Psi(R, \tau)$ we sample a new distribution: $f(R, \tau) = \Psi(R, \tau)\psi_T(R)$ → SE:

$$\frac{\partial f(R, \tau)}{\partial \tau} = \underbrace{\nabla\left(\frac{1}{2}\nabla - v_D(R)\right)f(R, \tau)}_{\text{diffusion process}} - \underbrace{(E_L(R) - E_T)f(R, \tau)}_{\text{branching process}}$$

- $v_D(R) = \frac{\nabla\psi_T}{\psi_T} = \nabla \ln \psi_T$ – drift velocity, drives walkers in the direction of increasing $|\psi_T|$ → walkers are driven away from unimportant regions where $|\psi_T|$ is low and towards regions where

$|\psi_T|$ is significant, and with variable strength → the concentration of walkers is greater in important regions. $v_D \to \infty, \psi_T \to 0$ – at nodal surface the drift velocity approaches infinity in the direction away from the surface → walkers are mostly prevented from crossing the nodes, but on rare occasions they do.

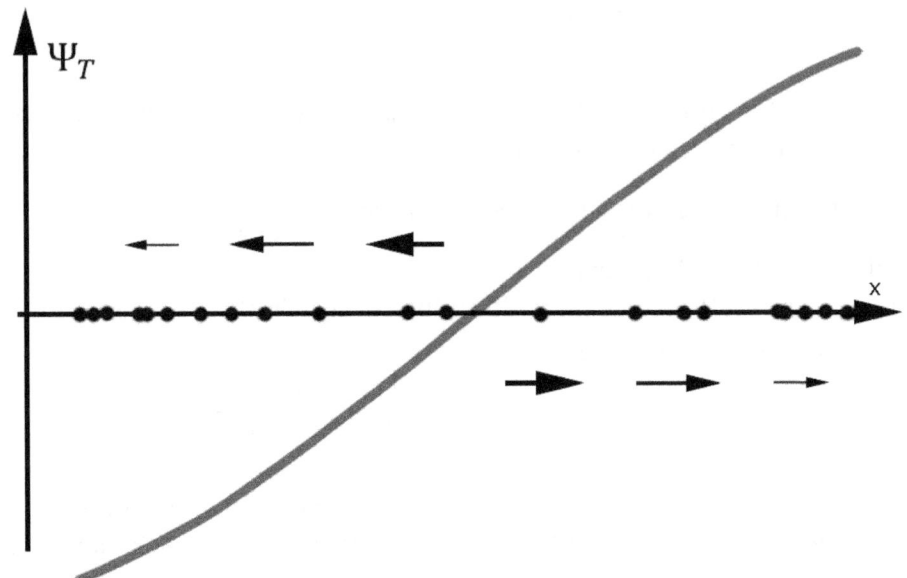

- $E_L(R) = \frac{\mathcal{H}\psi_T(R)}{\psi_T(R)}$ – local energy, a constant rate for the disappearance of walkers. A good ψ_T is close to ϕ_0 → E_L is close to E_0 → G_D does not vary greatly between steps and population fluctuations are reduced → importance sampling can be orders of magnitude more efficient than ordinary DMC.

- E_T is initially chosen to be VMC energy of ψ_T: $E_T = \frac{\langle \psi_T | \mathcal{H} | \psi_T \rangle}{\langle \psi_T | \psi_T \rangle}$, and is updated as the simulation progresses.

$f(R, \tau)$ has an asymptotic solution: $\lim\limits_{\tau \to \infty} f(R, \tau) = \psi_T(R) c_0 \phi_0(R) e^{-(E_0 - E_T)\tau}$

For importance sampling, in G_D we add drift velocity in the exponent, in G_B we replace potential with local energy and we get the short-time approximation to the Green's function:

- $G_D(R \to R', \Delta\tau) = (2\pi\Delta\tau)^{-\frac{3N}{2}} e^{-\frac{(R'-R)^2}{2\Delta\tau}} \to \tilde{G}_D(R \to R', \Delta\tau) = (2\pi\Delta\tau)^{-\frac{3N}{2}} e^{-\frac{(R'-R-v_D\Delta\tau)^2}{2\Delta\tau}}$ – $v_D = const.$ during $R \to R'$.

- $G_B(R \to R', \Delta\tau) = e^{-\frac{\Delta\tau}{2}(V(R)+V(R')-2E_T)} \to \tilde{G}_B(R \to R', \Delta\tau) = e^{-\frac{\Delta\tau}{2}(E_L(R)+E_L(R')-2E_T)}$ – a function of the surplus local energy.

9.2.4 Expectation values

Expectation values are obtained by averaging some function over a set of N samples of configuration space $\{R_n\}$ taken from the DMC run.

Multiply SE with trial function on the right: $\psi_T \mathcal{H}\Psi = \psi_T E\Psi \Rightarrow \int \psi_T \mathcal{H}\Psi dR = \int \psi_T E\Psi dR$

Wave functions are Hermitian $\rightarrow \int \Psi \mathcal{H}\psi_T dR = \int \Psi E\psi_T dR \Rightarrow \int \Psi \psi_T \frac{\mathcal{H}\psi_T}{\psi_T} dR = E \int \Psi \psi_T dR \Rightarrow$

$$E = \frac{\int \Psi \psi_T \frac{\mathcal{H}\psi_T}{\psi_T} dR}{\int \Psi \psi_T dR} = \frac{\int \Psi \psi_T E_L dR}{\int \Psi \psi_T dR} = \frac{\langle \Psi | \mathcal{H} | \psi_T \rangle}{\langle \psi_T | \psi_T \rangle} - \text{mixed estimator.}$$

For ground state energy the state begins as ψ_T and in the future it will become $\Psi \equiv \phi_0$: $\psi_T \rightarrow \phi_0$:

$$E_0 = \frac{\langle \phi_0 | \mathcal{H} | \psi_T \rangle}{\langle \phi_0 | \psi_T \rangle} = \frac{\int \phi_0 \mathcal{H}\psi_T dR}{\int \phi_0 \psi_T \, dR} = \lim_{\tau \to \infty} \frac{\int f E_L dR}{\int f dR} \approx \frac{1}{N} \sum_{n=1}^{N} E_L(R_n)$$

Expectations of quantities that do not commute with the Hamiltonian (S) can be calculated using a combination of the mixed and variational estimators:

$$\langle \phi | S | \phi \rangle \approx 2\langle \phi | S | \psi_T \rangle - \langle \psi_T | S | \psi_T \rangle + O((\phi - \psi_T)^2)$$

For nonnegative quantities:

$$\langle \phi | S | \phi \rangle \approx \frac{\langle \phi | S | \psi_T \rangle^2}{\langle \psi_T | S | \psi_T \rangle} + O((\phi - \psi_T)^2)$$

9.2.5 The Fermion problem

Review: Spin

Projection of spin of a composite particle is equal to the sum of projections of particle's spins from which it is made of:

$$s_{1+2} = |s_1 - s_2|, \dots, s_1 + s_2$$

$$(s_{1+2})_z = (s_1)_z + (s_2)_z$$

Notations (the last is spinor):

- $\left| \frac{1}{2}, \frac{1}{2} \right\rangle \equiv |\uparrow\rangle \equiv \begin{pmatrix} 1 \\ 0 \end{pmatrix}$
- $\left| \frac{1}{2}, -\frac{1}{2} \right\rangle \equiv |\downarrow\rangle \equiv \begin{pmatrix} 0 \\ 1 \end{pmatrix}$

For composite particle made from 2 fermions of spin ½ (e.g. 2 electrons or proton and electron in the ground state of hydrogen):

$$\left| \frac{1}{2} \right\rangle \otimes \left| \frac{1}{2} \right\rangle = \begin{cases} 0, s_z = 0 - \text{singlet} \\ 1, s_z = \{-1,0,1\} - \text{triplet} \end{cases}$$

- Singlet: $\frac{1}{\sqrt{2}}(|\uparrow\rangle|\downarrow\rangle - |\downarrow\rangle|\uparrow\rangle)$
- Triplet:
  - -1: $|\downarrow\rangle|\downarrow\rangle$
  - $+1$: $|\uparrow\rangle|\uparrow\rangle$

  o $0: \frac{1}{\sqrt{2}} \left(|\!\uparrow\rangle |\!\downarrow\rangle + |\!\downarrow\rangle |\!\uparrow\rangle \right)$

In spinor notation:

$$\binom{\alpha}{\beta} = \alpha \binom{1}{0} + \beta \binom{0}{1}\text{s}$$

- $|\alpha|^2$ – probability that the measurements of projection of spin will return $\frac{1}{2}\hbar$,
- $|\beta|^2$ – that we will get $-\frac{1}{2}\hbar$.

$$|\alpha|^2 + |\beta|^2 = 1$$

Let's measure the components of the spin: $\{s_x, s_y, s_z\}$:

$$s_x = \frac{\hbar}{2}\begin{pmatrix} 0 & 1 \\ 1 & 0 \end{pmatrix} = \frac{\hbar}{2}\sigma_x, s_y = \frac{\hbar}{2}\begin{pmatrix} 0 & -i \\ i & 0 \end{pmatrix} = \frac{\hbar}{2}\sigma_y, s_z = \frac{\hbar}{2}\begin{pmatrix} 1 & 0 \\ 0 & -1 \end{pmatrix} = \frac{\hbar}{2}\sigma_z$$

$\sigma_x, \sigma_y, \sigma_z$ – Pauli matrices

System of particles

Let x be a set of quantum # for single-particle state. Consider a system of 2 identical particles with states: x_1 and x_2. Because the particles are indistinguishable, either particle can be in either state – the system is in superposition of both states prior to measurement:

- Symmetric state: $|x_1, x_2\rangle \propto |x_1\rangle |x_2\rangle + |x_2\rangle |x_1\rangle$
- Antisymmetric state: $|x_1, x_2\rangle \propto |x_1\rangle |x_2\rangle - |x_2\rangle |x_1\rangle$

For a system of either bosons or fermions the wave function must have the correct properties of symmetry and anti-symmetry.

- System of bosons → wave function is positive and the positive density of walkers represents the positive wave function → no problem, since there are no nodes[22].
- System of two fermions of opposite spin for which the wave function is symmetric to the exchange of two electrons → also no problem.
- System of n fermions → the wave function must be antisymmetric to the exchange of fermions of the same spin → wave function goes negative in some regions, with nodes where the sign changes → a problem: how do we represent a negative wave function with a positive density of walkers?

---

[22] A node is a point along a standing wave where the wave has minimum amplitude. The nodal surface of a wave function $\Psi(R)$ in dN-dimensions is the $(dN - 1)$-dimensional surface on which $\Psi = 0$ and across which Ψ changes sign.

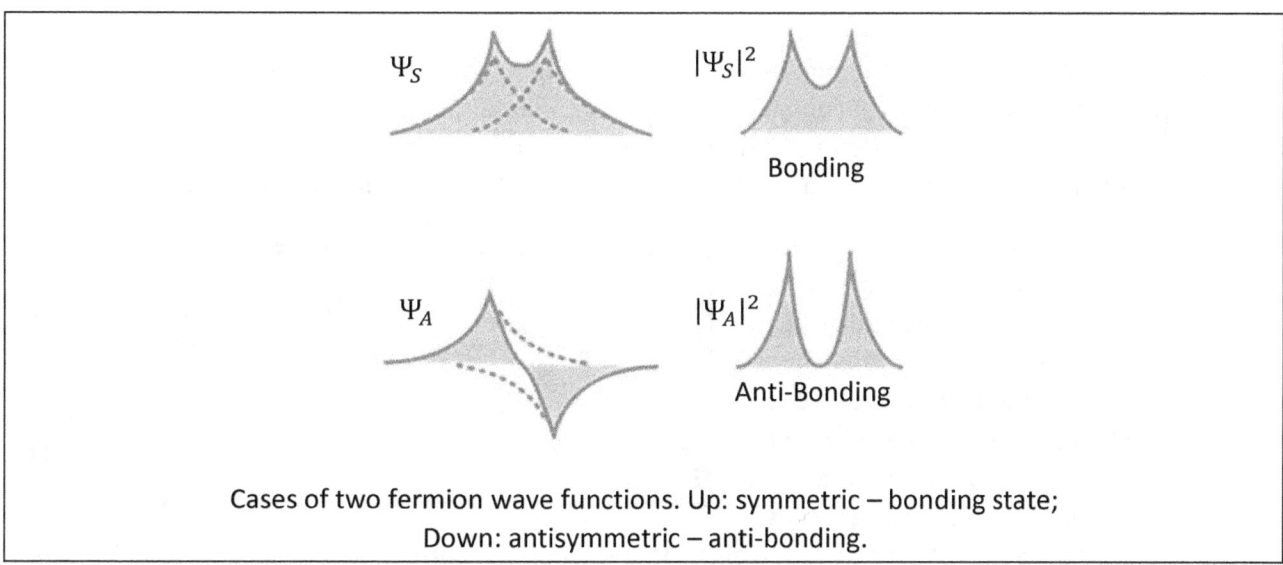

Cases of two fermion wave functions. Up: symmetric – bonding state;
Down: antisymmetric – anti-bonding.

Except in the simplest cases the wave function for a system of n fermions is positive and negative in different regions of the $3n$-dimensional space → the regions are separated by one or more $(3n - 1)$-dimensional hypersurfaces which cannot be specified except by the solution of SE.

System of two electrons of the same spin

The configuration space is divided in half by the nodal surface. The two halves are similar in shape and are nested together face-to-face. The positions of the two electrons are represented by a single point in configuration space and interchange of the two electrons moves the point across the nodal surface to a similar position in the other half of the configuration space.

Example: He atom

Case: $1s2s\ ^3S$ → the wave function may be regarded as a function of the electron-nucleus radii r_1, r_2 and the angle θ between them.

$$\Psi = 1s(1)2s(2) - 1s(2)2s(1) = 0, r_1 = r_2$$

The nodal surface is $(3 \cdot 2 - 1) = 5$-dimensional hypersurface on which $r_1 = r_2$.

The configuration space is divided by the nodal surface into two equivalent sections, one with positive, and one with negative wave function.

Case: $1s2s\ ^2P$ → the symmetry properties are not sufficient to specify the node structure because there are many possible wave functions which have the required anti-symmetry on reflection in the $z = 0$ plane and on exchange of electrons.

For such complex cases we need DMC.

The fixed-node approximation

A simple solution is to fix the position of nodes by taking them from ψ_T → the nodes of ψ_T define the nodal hypersurface, which divides configuration space into nodal pockets. For determining ψ_T and its nodes we can use VMC or even better, Hartree-Fock. The fixed-node DMC energy is an upper bound to the exact ground state energy. The results are not exact unless the trial nodal surface is exact.

The best example to introduce the fixed-node approximation is a particle in a box. Walkers, initially distributed according to the starting state (dashed line at the image), diffuse randomly until they cross one of the box walls, when they are removed from the simulation. The absorption at the walls ensures that the average walker density

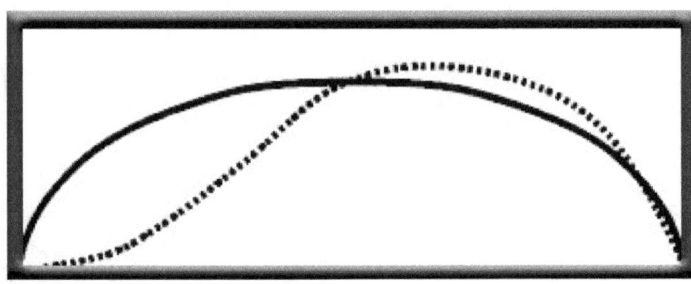

tends to zero at the edges of the box → the eigenstate satisfies the correct boundary conditions. E_T is set so that diffusing walkers slowly increase their numbers. Regardless of the shape of the initial distribution, the walker density settles down to the final state (full line on image) which is proportional to the ground-state wave function.

Example with first excited state of a particle in a box → odd parity → a node at the box center. One solution would be to start with a trial function of odd parity, and write it as the difference between the two non-negative distributions: $\Psi = \Psi_+ - \Psi_-$ [23]

The fixed-node solution: introducing one extra absorbing barrier at the center of the box → the space is divided into two separate simulation regions. After a long time, the walker densities are proportional to the lowest-energy eigenfunctions satisfying zero boundary conditions at both the box walls and the absorbing barrier → the $t \to \infty$ walker densities within each region must be proportional to the odd-parity eigenfunction in that region.

Nodal pockets

The many-electron ground state is very complicated, but the fixed-node approximation works the same way. If the nodes are known exactly, absorbing barriers may be places everywhere on the nodal surface, dividing up the configuration space into a set of disjoint nodal pockets. Parallel DMC simulations are then carried out in all nodal pockets → the solution within each pocket tends to ± the exact ground-state wave function in that pocket. Within each pocket, the fixed-node DMC projects out the lowest energy nodeless wave function satisfying zero boundary conditions on the enclosing nodal surface.

The only communication between the simulations in different pockets is via E_T, which is gradually adjusted to keep the total walker population roughly constant.

---

[23] The same principle will later be used in the release-node method.

- Ground state: by tiling theorem, the wave functions of the ground state are all equivalent by the symmetry and thus are equally favorable.
- Excited state: if some pockets are more favorable than others → the walker population becomes more concentrated there → energy tends to the eigenvalue of the lowest-energy nodeless wave function within the most favorable pockets.

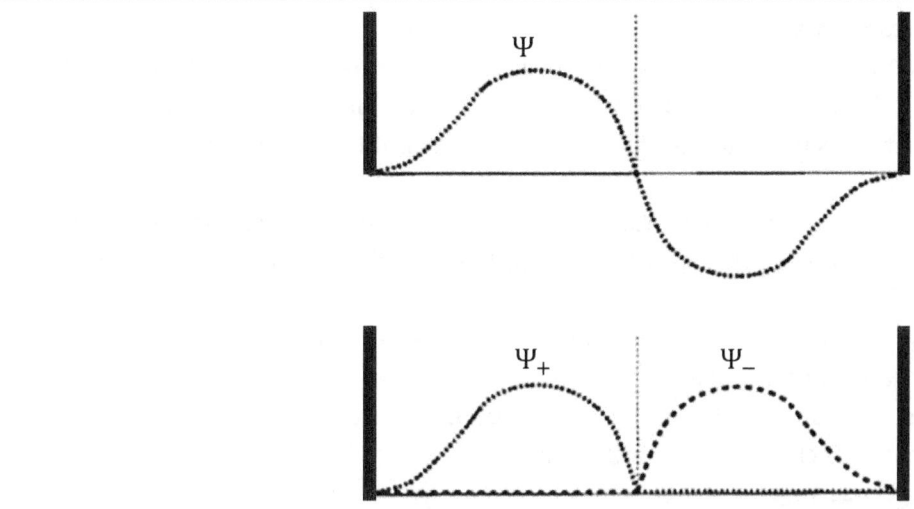

Example: The wave function $\Psi(R)$ of N electrons in 1D is equal to zero whenever any two electorns coincide → this defines $(N-1)$-dimensional coincidence planes passing through the N-dimensional configuration space.

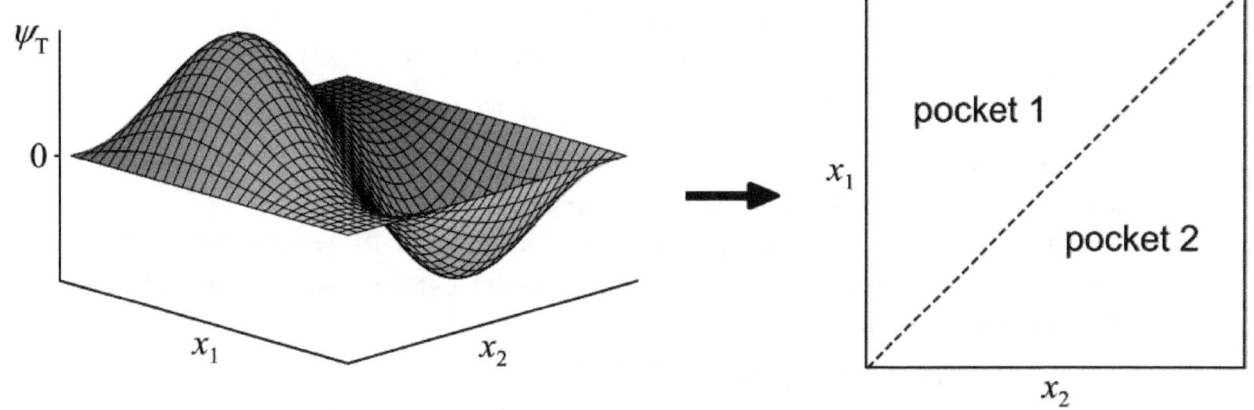

Example. 1D square well and two electrons. $\psi_T(x_1, x_2)$ can be constructed analytically from the case where the electrons do not interact, and used to define nodal pockets.

DMC is used to determine ϕ_0 in each pocket, using the boundary condition: $\psi_T(R) = 0 \Rightarrow \phi_0(R) = 0$

We enforce the condition $(\forall \tau) f(R, \tau) = 0$ → the nodes are fixed. This can be done if we make $\tilde{G}_D(R \to R', \Delta\tau) = 0$ when R and R' are in different pockets. For infinitely small $\Delta\tau$ this is always true, but for any finite time-step some walkers will cross nodal surfaces → reject crossings.

Algorithm for fixed-node DMC

1. Generate a set of walkers from some initial distribution (often from $|\psi_T|^2$). Initialize E_T to the average VMC energy of the ensemble.
2. For each walker:
    a. Make a trial move $R \to R': R' = R + \chi + v_D \Delta\tau$

    χ – 3N-vector of normally distributed random numbers with variance $\sigma = \Delta\tau$ and $\mu = 0$.

        i. **Rejectance step**

    Walkers on rare occasions attempt to cross the nodal surface → failure of the approximate Green's function to describe the region close to the node. Things get better as time-step tends to zero: $\Delta\tau \to 0$, but this makes calculation inefficient.

    We must eliminate the walker that crosses the node or reject the move and keep the walker at its original position. The rejection method gives smaller time-step error.

    Thus, if the walker has crossed a nodal hypersurface → the sign of the walker has changed → reject a walker.

    Cross-recross error: if the walker has crossed an even number of hypersurfaces it would not change a sign and would be accepted.

        ii. **Acceptance step**

    Changing configuration $R \to R'$:

          1. Without importance sampling:

    $$\Psi(R', \tau + \Delta\tau) = \int \Psi(R, \tau) G(R \to R', \Delta\tau) dR$$

          2. With importance sampling:
    $$f(R', \tau + \Delta\tau) = \Psi(R', \tau + \Delta\tau)\psi_T(R')$$
    $$= \int \Psi(R, \tau)\psi_T(R)\tilde{G}(R \to R', \Delta\tau) dR$$

Comparing 1. and 2.: $G(R \to R', \Delta\tau) = \psi_T^{-1}(R)\tilde{G}(R \to R', \Delta\tau)\psi_T$

G should be symmetric in R and R': $G(R \to R', \Delta\tau) = G(R' \to R, \Delta\tau)$

We get: $G(R' \to R, \Delta\tau) = \psi_T^{-1}(R)\tilde{G}(R' \to R, \Delta\tau)\psi_T == G(R \to R', \Delta\tau) = \psi_T^{-1}(R)\tilde{G}(R \to R', \Delta\tau)\psi_T \to \tilde{G}(R' \to R, \Delta\tau)\psi_T^2 = \tilde{G}(R \to R', \Delta\tau)\psi_T^2$ – the detiled balance equation for DMC with importance sampling.

Weight for the move: $W(R, R') = \dfrac{\tilde{G}(R' \to R, \Delta\tau)\psi_T^2(R')}{\tilde{G}(R \to R', \Delta\tau)\psi_T^2(R)}$

Thus, we accept with Metropolis probability:
$$A(R \to R') = \min(W(R, R'), 1)$$

This step ensures that we sample correctly for finite $\Delta\tau$, allowing the use of a larger $\Delta\tau$ than would otherwise be acceptable.

    b. Branching step
- $G_B < 1$ → the walker continues its evolution with probability G_B,
- $G_B \geq 1$ → the walker continues, in addition, at the same position, a new walker is created with probability $(G_B - 1)$.

Both possibilities can be coded as a single command M. The examined walker becomes M walkers positioned at the same point in configuration space as the original walker:

$$M = \text{int}\left(\eta + \tilde{G}_B(R \rightarrow R', \Delta\tau)\right) = \begin{cases} 0, \text{death of walker} \\ 1, \text{the walker proceeds further} \\ 2, \text{birth of a walker} \\ \vdots \end{cases}$$

$$\eta \sim U(0,1)$$

3. Quantities of interest are accumulated, weighted by the branching factor $\tilde{G}_B$.
4. Periodically adjust the total population of walkers via adjustment of E_T:

$$E_T \rightarrow E_T - c\ln\frac{M}{M_{\text{avg}}}$$

   M – current # of walkers, M_{avg} – target # of walkers, c – appropriate constant that controls how quickly the actual number of walkers, M, approaches the desired number M_{avg} (it should rebalance the # of walkers in 10-50 time-steps).
5. Repeat steps 2-4 for each walker until the system has equilibrated (until the fluctuations in the quantities of interest have become only statistical).
6. Reset accumulators and continue the simulation, extracting the final results.

Improving the algorithm

Reducing time-step error

For simple atoms and molecules the time-step error can be reduced up to 300 times, by making several improvements:

- Using the effective time-step $\Delta\tau_{\text{eff}}$, which varies with the rate at which walkers are currently diffusing. As the drift velocity diverges on the nodal surface → walker close to a node can make a quite large move. The problem can be solved by imposing a cutoff on the magnitude of the drift velocity and on the values of the local energy.
- The branching factor is modified to depend also on acceptance probability → reduces fluctuations in the total number of walkers.
- Instead of creating of random numbers of walkers in the branching step, walkers are deleted or combined pairwise → also reduces fluctuations.
- The mixed estimator for energy $E^{\text{mix}} = \frac{\langle\Psi|\mathcal{H}|\psi_T\rangle}{\langle\psi_T|\psi_T\rangle}$ is modified to depend on $E_L(R), E_L(R'), A(R \rightarrow R')$ → reduces the variance in E^{mix}.
- v_D diverges at the nodes and is discontinuous at nuclei → we can modify Green's function to include this into account.

Zero-variance algorithm

This method is to reduce the computational cost by approximately 10 times. Algorithm:

1. Renormalize $E_L(R)$ so that it has the same mean, but a lower variance.
2. As the length of the MC run tends to ∞ → variance in the observable tends to 0 → optimize the renormalized $E_L(R)$ over a short MC run – zero-variance condition.
3. Perform a long MC using renormalized $E_L(R)$.

Improving on Fixed-Node

There are several methods:

1. Release-node: proceed as for the fixed-node, but then remove the fixed-node constraint – information is extracted as the system relaxes into a nodeless Bose-like state.
2. Exact cancelation: we label the walkers positive and negative and allow them to annihilate on contact, removing the need for predefined nodes.
3. Proper understanding of nodal surfaces: if it were possible to know the exact nodal surfaces, the fixed-node method would be exact.

Release node method

The release-node method is exact but expensive, and with another drawback of being restricted to systems where either:

i. the trial nodes are accurately located so that the relaxation to the true nodes occurs quickly, or
ii. the Pauli principle is unimportant so that $(E_F - E_B)$ is small. E_F – fermion ground state energy, E_B - boson ground state energy.

The release node or transient estimate methods proceed as for the fixed-node approach, but then we remove the fixed-node constraint. Information is extracted as the system relaxes into a nodeless Bose-like state.

A fermionic wave function can be decomposed into positive and negative function:

$$\Psi(R,\tau) = \Psi_+(R,\tau) - \Psi_-(R,\tau), \Psi_\pm = \frac{1}{2}(|\Psi| \pm \Psi)$$

$\tilde{G}$ acts on $\Psi_\pm$ independently → DMC may be applied to symmetric and antisymmetric parts of the wave function separately. No fixed-node constraint → the two functions will forget that they are separate parts of a fermion wave function → they will tend towards a boson ground state → the correct fermionic wave function can be extracted from the difference between the two simulations.

$$\Psi(R',\tau + \Delta\tau) = \int \Psi(R,\tau)\tilde{G}(R \to R',\Delta\tau)dR$$
$$= \int \Psi_+(R,\tau)\tilde{G}(R \to R',\Delta\tau)dR - \int \Psi_-(R,\tau)\tilde{G}(R \to R',\Delta\tau)dR$$
$$= \Psi_+(R',\tau + \Delta\tau) - \Psi_-(R',\tau + \Delta\tau)$$

Algorithm:

i. $t \in [0, \tau + \Delta\tau)$: Nodes are fixed: start by $\Psi(R,\tau)$ and by fixed-node calculate $\Psi_\pm = \frac{1}{2}(|\Psi| \pm \Psi)$
ii. $t \geq \tau + \Delta\tau$: Nodes are released: $\Psi_\pm$ tend to their bosonic-ground states → $\Psi_\pm(R',\tau + \Delta\tau)$
iii. $\Psi(R',\tau + \Delta\tau) = \Psi_+(R',\tau + \Delta\tau) - \Psi_-(R',\tau + \Delta\tau)$

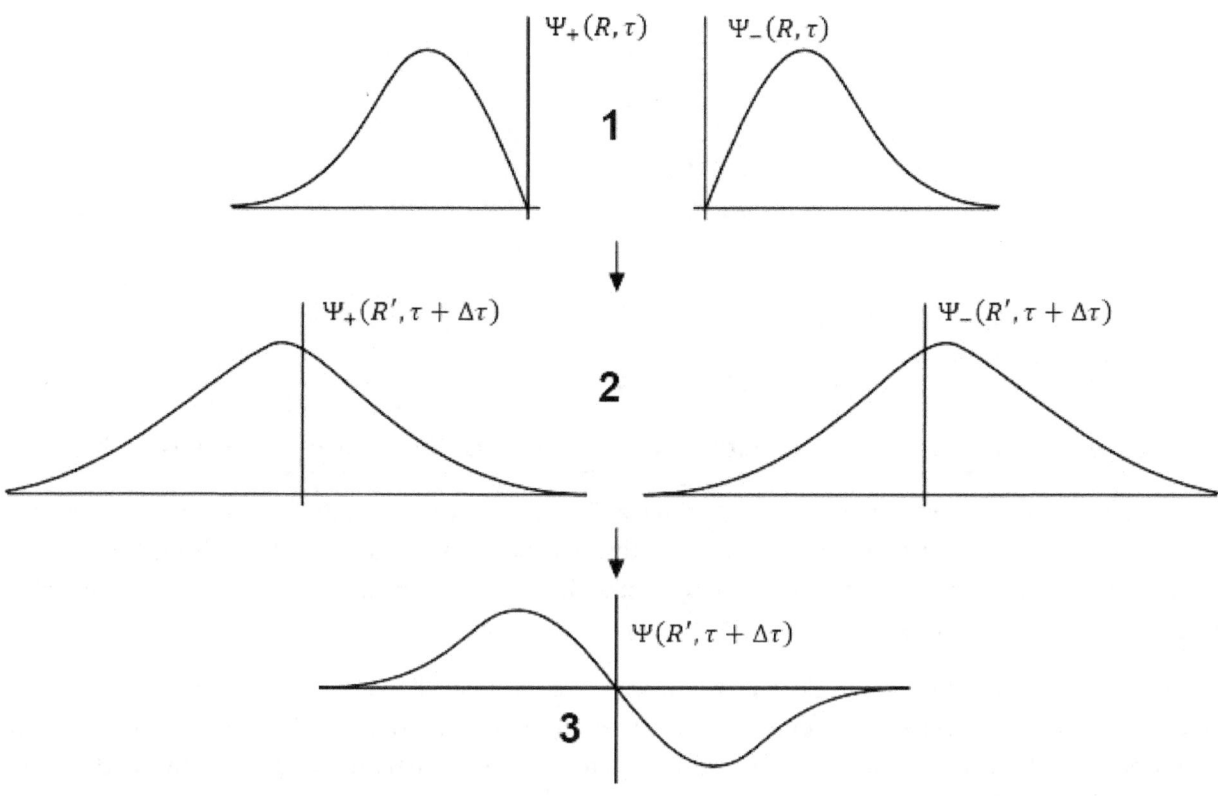

After the nodes are released the bosonic contribution starts to dominate the fermionic contribution → soon after the nodes are released the fermion component will become swamped in statistical noise. In the time between the release and the decrease of fermionic component to noise useful information can be extracted from the simulation: a transient estimate. Transient energy estimator:

$$E_0^{\text{trans}} = \frac{\langle \Psi(R', \tau + \Delta\tau) | \mathcal{H} | \Psi(R, \tau) \rangle}{\langle \Psi(R', \tau + \Delta\tau) | \Psi(R, \tau) \rangle}$$

Exact cancelation

Properties:

- Combines the best features of the fixed- and release-node methods,
- Overcomes the node problem for small systems → provides exact solutions,
- Successfully applied to systems such as: $H_3, He_2, HeH, He_3,$
- It takes full advantage of the symmetric and antisymmetric properties of wave functions,
- It offers pairwise, self, and multiple collective cancellations of walkers.

In further text we will investigate an example: 1D LHO, 1st excited state.

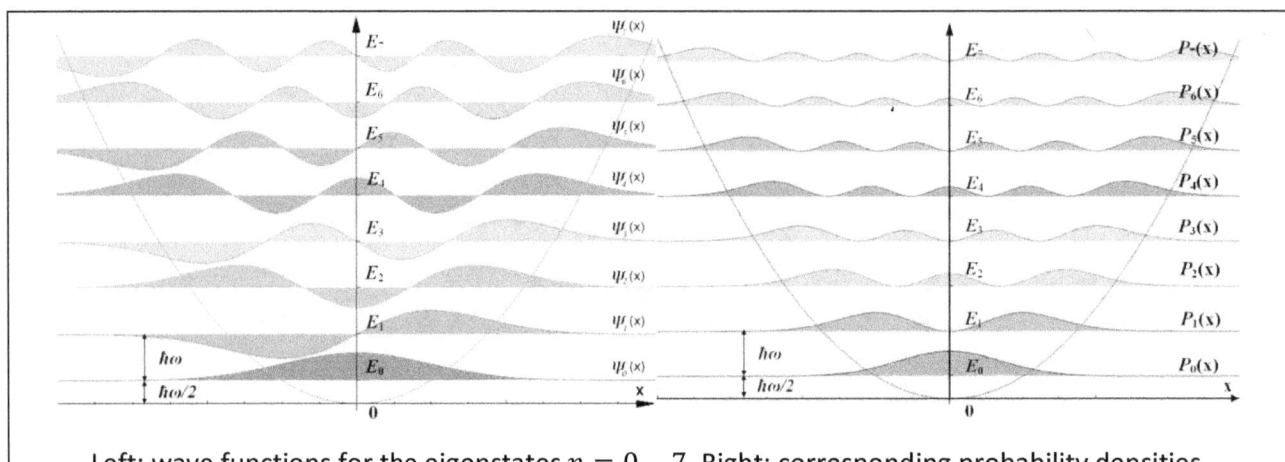

Left: wave functions for the eigenstates $n = 0 \dots 7$. Right: corresponding probability densities.

A QMC calculation can be carried out with positive and negative walkers, initially separated left and right from the center. In the absence of cancellation, the two populations spread throughout the available configuration space → penetrate each other → approach the symmetric distribution for the ground state, not the excited state we desire.

If positive and negative walkers in close proximity are occasionally allowed to cancel each other, the two populations tend to cancel each other and produce separated distributions, which reflects the one for the 1st excited state, if the two populations are controlled to maintain equal numbers of positive and negative walkers and if cancellations are properly executed.

One method of cancellation that combines accuracy with speed is cancelation on the basis of the overlap of the distributions (Green's functions) to which the walkers are moved.

Algorithm for 2 walkers:

1. Assign each walker a positive weight W and a sign s: $W_{1,2}, s_{1,2}$, at positions R_1 and R_2.

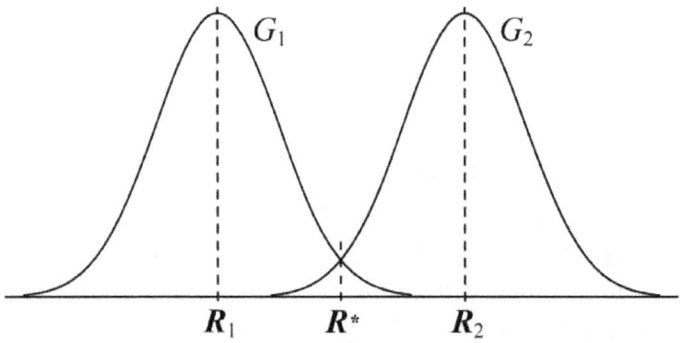

2. The walkers move to a new positions: $R_1 \to R_1', R_2 \to R_2'$ → Green's functions are:
$$G_{1,2}\left(R_{1,2} \to R_{1,2}', \tau + \Delta\tau\right)$$

3. Re-weight the walkers: $W_1' = \frac{\max(s_1 W_1 G_1 - s_2 W_2 G_2, 0)}{s_1 G_1}, W_2' = \frac{\max(s_2 W_2 G_2 - s_1 W_1 G_1, 0)}{s_2 G_2}$ → if two walkers are of opposite sign and equal weight at the same position they will be canceled, if they are well separated, their weights will remain unchanged. Between the well-separated and at the same position: the weight of each walker is adjusted in relation to the proximity and weight of the other.

The overlap is by amount: $O_{\text{lap}} = \int \min(W_1 G_1, W_2 G_2)\, dR$

113

The procedure divides configuration space into two regions - cells, with:

1) $W_1 G_1 > W_2 G_2$
2) $W_1 G_1 < W_2 G_2$

Extension to n number of walkers divides configuration space into n Voronoi cells[24], with the ith cell having the greatest value of $W_i G_i$:

$$W_j' = \begin{cases} \dfrac{\sum_{i=1}^n s_i W_i G_i}{s_j G_j}, W_j G_j > W_i G_i \\ 0, W_j G_j < W_i G_i \end{cases}, i \neq j$$

Since there are multiple steady-state solutions for the ground state, fluctuations can shift the system from one solution to the other. In the case of the 1D LHO one solution is left-positive/right-negative, the other is left-negative/right-positive. We must prevent shifts between different states from occurring.

Cancelations must occur often enough to maintain an adequate ratio of positive to negative walkers in both positive and negative regions.

Stability requires a high ratio of positive to negative walkers in positive regions; negative to positive in negative regions.

The required number of walkers and the required cancellation rate depend on the system – some systems are more stable than others.

Nodal properties

Nodal surfaces can be very complex objects. On the image below is the example of VMC calculations performed on a 2D system of 161 spin-up electrons → $(2N - 1) = 321$-dimensional nodal hypersurface. The image is a 2D slice (cross-section), allowing one electron to move (black dot), while others are held still (white dots).

---

[24] Without a formal definition, the Voronoi cells are self-explained by the following image:

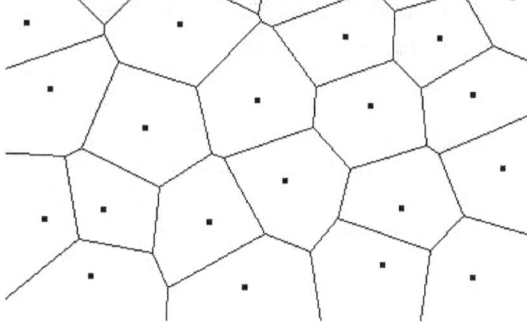

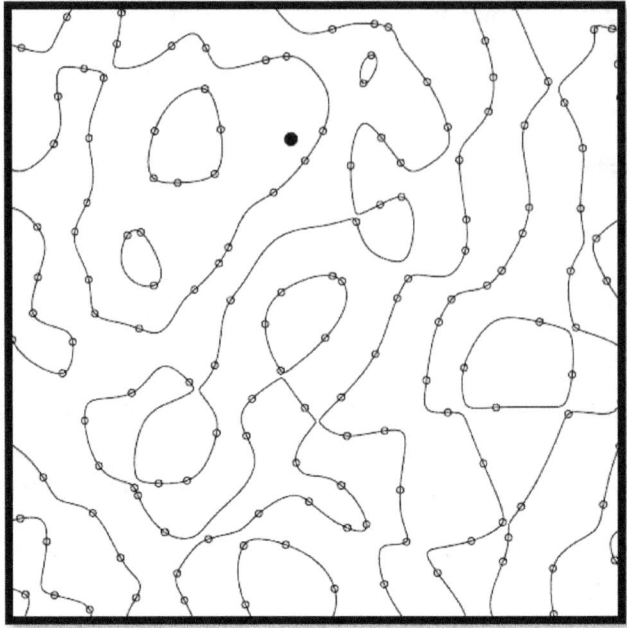

The facts about nodal surfaces:

- The exact nodal surface gives the lowest ground-state energy attainable, and any fixed-node simulation using any trial nodal hypersurface will produce a ground-state energy equal or greater than the exact energy – fixed node variational principle.
- The Pauli principle or the fact that the wave function will change sign if a pair of coordinates in the configuration are exchanged give us a set of coincidence[25] planes – the symmetry subsurface. Full nodal hypersurface is $(dN - 1)$-dimensional, while symmetry subsurface is $d(N - 1)$-dimensional → for $d = 1$ the nodal structure is fully determined by the symmetry arguments, but for $d > 1$ such arguments can provide only a points through which the nodes must pass.
- We may divide nodal pockets into classes with a permutation operator $\mathcal{P}$, i.e. group the pockets into classes equivalent by permutation symmetry:
    - i. Pick any point $X = (R, \sigma)$ in any nodal pocket and label that pocket A.
    - ii. Apply the transformation: $X \to \mathcal{P}X$.
    - iii. If the new X is in a new nodal pocket, label that nodal pocket A also.
    - iv. Repeat steps ii. and iii. until all permutations in $\mathcal{P}$ are exhausted.
    - v. In the next unlabeled nodal pocket, pick a point and label a pocket B.
    - vi. Repeat steps ii.-iv. for the pocket B.
    - vii. Repeat the procedure until all nodal pockets have been labeled.
- The tilling theorem: if we have a local potential all nodal pockets in the ground state of many-fermion wave function will belong to the same class.

---

[25] *Coincident* = they are one and the same.

The fixed-node variational principle

Pockets are divided into equivalence classes and Ψ_T is used to define the nodal surface for a fixed-node DMC → walkers will eventually concentrate in the nodal pockets of the lowest-energy class → the walker distribution will tend to the pocket ground state $\Psi_0^\alpha(R)$ – real, normalized wave function that is zero outside Ω^α-pocket and satisfies the fixed-node boundary conditions. Within its pocket the ground-state is a nodeless eigenfunction of the SE with eigenvalue E_0^α, while outside it is zero:

$$\begin{cases} \mathcal{H}\Psi_0^\alpha(R) = E_0^\alpha \Psi_0^\alpha(R) + \delta^\alpha, R \in \Omega^\alpha \\ \Psi_0^\alpha(R) = 0, R \notin \Omega^\alpha \end{cases}$$

δ – occurs on the nodal surfaces.

The DMC energy for pocket Ω^α is equal to E_0^α → any symmetry equivalent pocket will have the same energy.

- $\mathcal{P}$ – permutation operator,
- $N_\mathcal{P}$ - # of permutations.

Define the Hermitian, antisymmetrization operator $\hat{A}$, acting on a function $f(R)$:

$$\hat{A}f(R) = \frac{1}{N_\mathcal{P}} \sum_\mathcal{P} (-1)^\mathcal{P} f(\mathcal{P}R)$$

It is also idempotent: $\hat{A}\left(\hat{A}f(R)\right) = \hat{A}f(R)$

We use $\hat{A}$ on Ψ_0^α to construct a real antisymmetric wave function: $\widetilde{\Psi}_0^\alpha(R) = \hat{A}\Psi_0^\alpha$

Now we can use variational principle to show that the system's lowest energy eigenfunction is always less than or equal to the lowest energy eigenfunction of a nodal pocket:

$$E_0 \leq \frac{\langle \widetilde{\Psi}_0^\alpha | \mathcal{H} | \widetilde{\Psi}_0^\alpha \rangle}{\langle \widetilde{\Psi}_0^\alpha | \widetilde{\Psi}_0^\alpha \rangle} = \frac{\langle \widetilde{\Psi}_0^\alpha | \mathcal{H} | \hat{A}\Psi_0^\alpha \rangle}{\langle \widetilde{\Psi}_0^\alpha | \hat{A}\Psi_0^\alpha \rangle} = \frac{\langle \widetilde{\Psi}_0^\alpha \hat{A} | \mathcal{H} | \Psi_0^\alpha \rangle}{\langle \widetilde{\Psi}_0^\alpha \hat{A} | \Psi_0^\alpha \rangle} = \frac{\langle \hat{A}\widetilde{\Psi}_0^\alpha | \mathcal{H} | \Psi_0^\alpha \rangle}{\langle \hat{A}\widetilde{\Psi}_0^\alpha | \Psi_0^\alpha \rangle} = \frac{\langle \widetilde{\Psi}_0^\alpha | \mathcal{H} | \Psi_0^\alpha \rangle}{\langle \widetilde{\Psi}_0^\alpha | \Psi_0^\alpha \rangle} = E_0^\alpha$$

If the nodes of Ψ_T are exact → $E_0 = E_0^\alpha$ – the exact ground state cannot have more than one class of nodal pockets; otherwise $E_0 < E_0^\alpha$.

9.2.6 DMC with a complex wave function[26]

So far we have assumed that $\phi_0(R)$ and hence $f(R,\tau)$ are everywhere real. For isolated system this is not a problem: even if $\phi_0(R)$ is complex, the real and imaginary parts separately satisfy the SE → $\mathcal{H}$ observes time-reversal symmetry.

In the presence of an external field, for example a magnetic field, the time-reversal symmetry is broken → instead of the fixed-node approximation we must use the fixed phase approximation, and as the probability weight we take the module of the complex distribution function:

$$\phi_0(R) = |\phi_0(R)|e^{i\varphi(R)}$$

---

[26] Here is just the outline of the old method. The better method for complex DMC is given in "Complex Diffusion Monte-Carlo method for the systems with complex wave function…", 2013., B. Abdullaev at al.

All quantities are calculated by averaging over this complex distribution function.

It is possible to decompose the SE into 2 coupled differential equations in $|\phi(R)|$ only. The sign of $|\phi_0(R)|$ is fixed → DMC is used to solve the problem exactly → a choice is made for φ → $E_0 = E_0(\varphi)$ → variational method, with trial function φ_T, is used to minimize E_0.

9.3 Applications
Here we briefly present only a few basic examples of what can be done with VMC and DMC.

9.3.1 VMC

$$P(R) = \frac{|\psi_T(R)|^2}{\int |\psi_T(R)|^2 dR}$$

$$\langle E \rangle = \int P(R) E_L(R) dR$$

1D LHO

$$\mathcal{H} = -\frac{d^2}{dx^2} + x^2 \Rightarrow E_0 = 1$$

The exact ground-state wave function: $\psi_0 = \pi^{-1/4} e^{-\frac{x^2}{2}} \rightarrow \psi_T = \sqrt{\alpha}\pi^{-1/4} e^{-\alpha^2 \frac{x^2}{2}}$

$$E_L = \alpha^2 + x^2(1 - \alpha^4)$$

$$\langle E \rangle = \int_{-\infty}^{\infty} |\psi_T|^2 E_L dx = \left| \int_{-\infty}^{\infty} e^{-\alpha^2 x^2} dx \right| = \sqrt{\left|\frac{\pi}{\alpha}\right|} = \frac{\alpha^2}{2} + \frac{1}{2\alpha^2}$$

In solving the problem we can chose whether we wish to use Metropolis and sample over relevant configurations, or just use PRNG from a normal distribution, since the LHO wave functions follow closely such a distribution.

Hydrogen atom
Radial SE:

$$-\frac{\hbar^2}{2m}\frac{\partial^2 \mathcal{R}(r)}{\partial r^2} - \underbrace{\left(k\frac{e^2}{r} - \frac{\hbar^2}{2m}\frac{l(l+1)}{r^2}\right)}_{V_{\text{eff}}} \mathcal{R}(r) = E\mathcal{R}(r)$$

It is convenient to rewrite the equation in terms of dimensionless variables: $\rho = \frac{r}{\beta}$, $\beta = $ const.

$$-\frac{1}{2}\frac{\partial^2 \mathcal{R}(\rho)}{\partial \rho^2} - \left(\frac{mke^2\beta}{\hbar^2}\frac{1}{\rho} - \frac{l(l+1)}{2\rho^2}\right)\mathcal{R}(\rho) = \frac{m\beta^2}{\hbar^2}E\mathcal{R}(\rho)$$

We can determine β by requiring: $\frac{mke^2\beta}{\hbar^2} = 1 \Rightarrow \beta = \frac{\hbar^2}{mke^2} = a_0 = .05 \text{ nm} - \text{Bohr radius}.$

Introduce variable λ: $\lambda = \frac{m\beta^2}{\hbar^2}E$

$E = \frac{E_0}{n^2}, E_0 = -13.6$ eV, n – principal quantum # $\rightarrow \lambda = -\frac{1}{2n^2}$

SE now becomes:

$$\underbrace{-\frac{1}{2}\frac{\partial^2 \mathcal{R}(\rho)}{\partial \rho^2} - \left(\frac{1}{\rho} - \frac{l(l+1)}{2\rho^2}\right)\mathcal{R}(\rho)}_{\mathcal{H}\mathcal{R}} = \lambda \mathcal{R}(\rho)$$

The ground-state: $\lambda = -\frac{1}{2} \Leftrightarrow E_0 = -13.6$ eV. The exact wave function is: $\mathcal{R} = \rho e^{-\rho}$ $\rightarrow$ as a trial wave function we can use (α – parameter):

$$\mathcal{R}_T = \alpha \rho e^{-\alpha \rho}$$

$$E_L = \frac{\mathcal{H}\mathcal{R}_T}{\mathcal{R}_T} = -\frac{1}{2\mathcal{R}_T}\frac{\partial^2 \mathcal{R}(\rho)}{\partial \rho^2} - \left(\frac{1}{\rho} - \frac{l(l+1)}{2\rho^2}\right) = -\frac{1}{\rho} - \frac{\alpha}{2}\left(\alpha - \frac{2}{\rho}\right)$$

$$\langle E \rangle = \int P(R)E_L(R)dR = \int_0^\infty \alpha^2 \rho^2 e^{-2\alpha\rho} E_L(\rho)\rho^2 d\rho$$

We can use simple MC here, and we can make use of the density function, like in 3D torus example: set density to correspond to the exponential PDF. Even better is to use Metropolis algorithm.

Nucleon in a Gaussian potential

We observe a nucleus only. The nuclear interaction is short ranged ($\sim$1 fm). Pick out one nucleon, either proton or neutron, and we approximate the interaction between that lonely nucleon and the remaining nucleus with a Gaussian potential:

$$V(r) = V_0 e^{-\frac{r^2}{a^2}}, V_0 = -45 \text{ MeV}, a = 2 \text{ fm}$$

The average mass of one nucleon: $m = \frac{m_p + m_n}{2} \approx 938.926 \frac{\text{MeV}}{c^2} \rightarrow \frac{\hbar^2}{m} = 41.466 \text{ MeVfm}^2$

Assume that the nucleon is in $1s$ state and approximate the wave function of that LHO in the ground state:

$$\psi_T(r) = \alpha^{\frac{3}{2}}\pi^{-\frac{3}{4}}e^{-\frac{\alpha^2 r^2}{2}} \Rightarrow E_L = \frac{\hbar^2}{2m}(3\alpha^2 - r^2\alpha^4) + V(r)$$

$$\langle E \rangle = \int P(R)E_L(R)dR = \frac{3\hbar^2}{4m}\alpha^2 + V_0\left(\frac{a^2\alpha^2}{1 + a^2\alpha^2}\right)^{\frac{3}{2}}$$

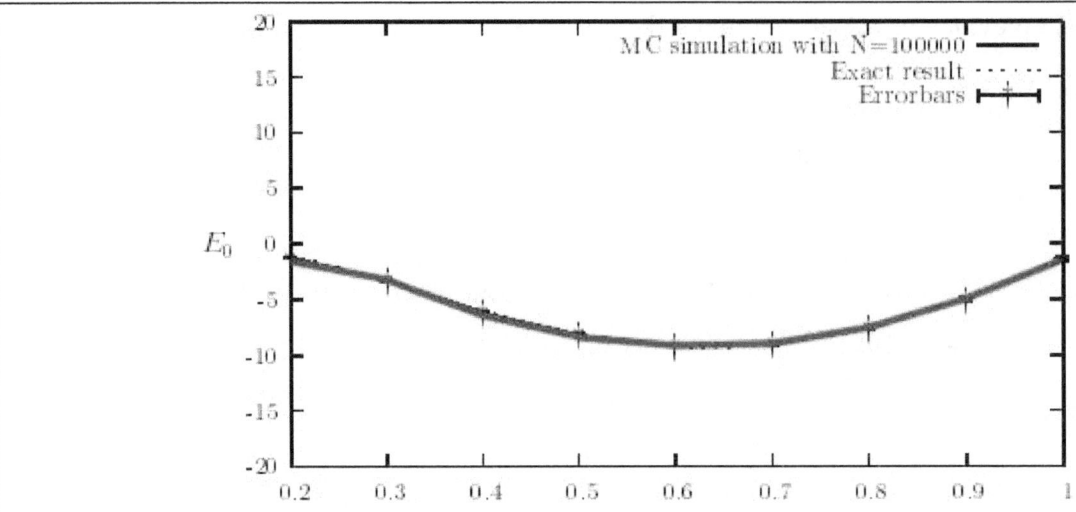

Results for the ground state energy of a nucleon in a Gaussian potential, comparing exact with the results from MC.

A system of N particles; Example: He

For a system of N particles:

$$\langle E \rangle = \frac{\int \Psi^*(R_1, \ldots, R_N)\mathcal{H}\Psi(R_1, \ldots, R_N)dR_1 \ldots dR_N}{\int \Psi^*(R_1, \ldots, R_N)\Psi(R_1, \ldots, R_N)dR_1 \ldots dR_N}$$

We use Born-Oppenheimer approximation.

For *He* atom:

$$V = -\frac{2ke^2}{r_1} - \frac{2ke^2}{r_2} + \frac{ke^2}{r_{12}} \Rightarrow \mathcal{H} = -\frac{\hbar^2}{2m}(\nabla_1^2 + \nabla_2^2) - \frac{2ke^2}{r_1} - \frac{2ke^2}{r_2} + \frac{ke^2}{r_{12}}$$

Approximation that we omit the repulsion between electrons has advantage that they can be treated as independent of each other – each electron only sees central field, but gives ground state energy of -108.8 eV, while the experimental value is -78.8 eV.

We need to choose a good trial wave function.

Cusp condition 1

Let us single out one electron and assume that it is close to the nucleus, i.e. $r_1 \to 0$. We also assume that the two electrons are far from each other, and that it is $l = 0$:

$$E_L(R) = \frac{1}{\psi_T}\left(-\frac{1}{2}\nabla_1^2 - \frac{Z}{r_1}\right)\psi_T + \text{finite terms} = \frac{1}{\mathcal{R}_T(r_1)}\left(-\frac{1}{2}\frac{d^2}{dr_1^2} - \frac{1}{r_1}\frac{d}{dr_1} - \frac{Z}{r_1}\right)\mathcal{R}_T(r_1) + \text{finite terms}$$

$$\lim_{r_1 \to \infty} E_L = \frac{1}{\mathcal{R}_T(r_1)}\left(-\frac{1}{r_1}\frac{d}{dr_1} - \frac{Z}{r_1}\right)\mathcal{R}_T(r_1) = 0 \Rightarrow \frac{1}{\mathcal{R}_T}\frac{d\mathcal{R}_T}{dr_1} = -Z \Rightarrow \mathcal{R}_T(r_1) \propto e^{-Zr_1}$$

Similar for second electron: $\mathcal{R}_T(r_2) \propto e^{-Zr_2}$

For $l > 0$: $\frac{1}{\mathcal{R}_T}\frac{d\mathcal{R}_T}{dr} = -\frac{Z}{l+1}$

119

Cusp condition 2

When two electrons approach each other → the wave function dependence on r_{ij} should reflect the correct behavior when $r_{ij} \to 0$. The resulting radial equation is the same as for the electron-nucleus case, except that the attractive Coulomb interaction is replaced by a repulsive interaction, and the kinetic energy term is twice as large:

$$\lim_{r_{ij}\to 0} E_L(R) = \frac{1}{\mathcal{R}_T(r_{ij})}\left(-\frac{4}{r_{ij}}\frac{d}{dr_{ij}} + \frac{2}{r_{ij}}\right)\mathcal{R}_T(r_{ij}) = 0 \Rightarrow \begin{cases} \dfrac{1}{\mathcal{R}_T}\dfrac{d\mathcal{R}_T}{dr_{ij}} = \dfrac{1}{2}, l = 0 \\ \dfrac{1}{\mathcal{R}_T}\dfrac{d\mathcal{R}_T}{dr_{ij}} = \dfrac{1}{2(l+1)}, l = 0 \end{cases}$$

Here we got another cusp condition.

For He atom: $r_{ij} \equiv r_{12}$

A possible trial wave function which reflects the cusp-condition between two electrons:

$$\psi_T = e^{-\alpha(r_1+r_2)+\frac{r_{12}}{2}}, \alpha - \text{variational parameter, effective charge}$$

H_2^+

- Proton 1 is at position $x = -\frac{R}{2}$,
- proton 2 is at $x = \frac{R}{2}$,
- distance of electron from the coordinate system is r,
- proton 1 – electron distance: $r_1 = \left|r - \frac{R}{2}\right|$,
- proton 2 – electron distance: $r_2 = \left|r + \frac{R}{2}\right|$.

Approximations:

i.   protons do not move,
ii. omit contributions from nuclear forces, since they act at distances of several orders of magnitude smaller than the equilibrium position.

SE:

$$\left(\underbrace{-\frac{\hbar^2}{2m}\nabla_r^2}_{\substack{\text{Kinetic energy}\\\text{of electron}}} \underbrace{-\frac{ke^2}{r_1} - \frac{ke^2}{r_2}}_{\substack{\text{Potential energy}\\\text{of electron}}} + \underbrace{\frac{ke^2}{R}}_{\text{p-p repulsion}}\right)\psi(r,R) = E\psi(r,R)$$

$$\underbrace{\phantom{-\frac{ke^2}{r_1} - \frac{ke^2}{r_2} + \frac{ke^2}{R}}}_{V(r,R)}$$

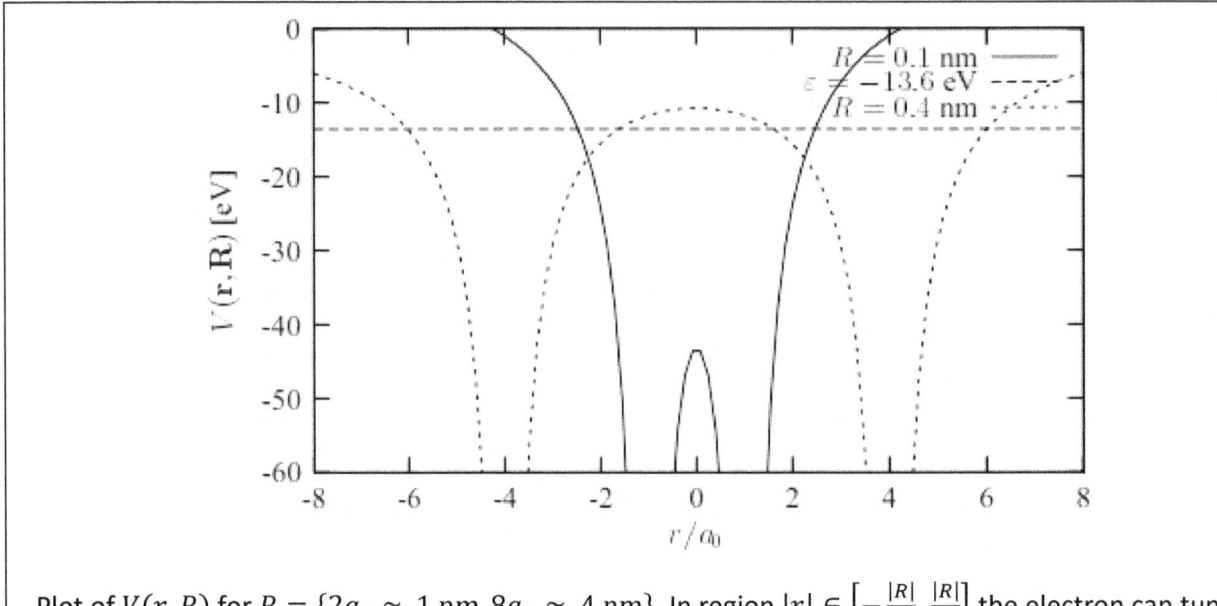

Plot of $V(r, R)$ for $R = \{2a_0 \approx .1\,\text{nm}, 8a_0 \approx .4\,\text{nm}\}$. In region $|r| \in \left[-\frac{|R|}{2}, \frac{|R|}{2}\right]$ the electron can tunnel through the potential barrier.

Since the potential is symmetric → the probability for electron to move from one proton to the other must be equal in both directions → the electron shares its time between protons.

If $R \rightarrow \infty$ → the electron is bound to one proton → hydrogen atom → as a trial wave function we can use the electronic wave function for the ground state of hydrogen: $\psi_{100} = \sqrt{\frac{1}{\pi a_0^3}} e^{-\frac{r}{a_0}}$

Since we do not know where the electron → we must allow for the possibility that the electron can be couplet do one of the protons → two hydrogen wave functions and their linear combination:

$$\psi_1 = \sqrt{\frac{1}{\pi a_0^3}} e^{-\frac{r_1}{a_0}}, \psi_2 = \sqrt{\frac{1}{\pi a_0^3}} e^{-\frac{r_2}{a_0}} \Rightarrow \psi_\pm = C(\psi_1 \pm \psi_2)$$

9.3.2 DMC

Solids

The non-relativistic Born-Oppenheimer Hamiltonian: $\mathcal{H} = -\frac{1}{2}\sum_i \nabla_i^2 - \sum_i \sum_\alpha \frac{Z_\alpha}{r_{i\alpha}} + \frac{1}{2}\sum_i \sum_{j \neq i} \frac{1}{r_{ij}}$

- r_i – electron position,
- $r_{i\alpha}$ – electron-nucleus distance,
- r_{ij} – electron-electron distance,
- Z_α – nucleus charge.

Phases of electron gas

- $n = \frac{1}{\frac{4\pi r_s^3}{3}}$ – number density

- $r_i' = \dfrac{r_i}{r_s}$ – scaled variables

$$\mathcal{H} = -\frac{1}{2r_s^2}\sum_{i=1}^{N}\nabla_{r_i'}^2 + \frac{1}{2r_s}\sum_{i=1}^{N}\sum_{\substack{j\neq i \\ j=1}}^{N}\frac{1}{r_{ij}'}$$

Cases:

i.   High-density (small r_s) limit: the interactions become negligible and the wave function may be approximated by HF determinant.

ii. Low-density (large r_s) limit: the Coulomb interactions dominate → kinetic energy can be ignored. The electrons behave like classical charges at zero temperature and should freeze into a Wigner crystal → the electron gas must undergo a first-order phase transition[27].

iii. Intermediate densities: the electron gas may become partially or wholly ferromagnetic.

DMC calculations require fine comparisons of the energies of phases with different trial wave functions, symmetries, and finite-size errors. As the density gets lowered, phase transitions can be observed: paramagnetic → ferromagnetic → Wigner crystal.

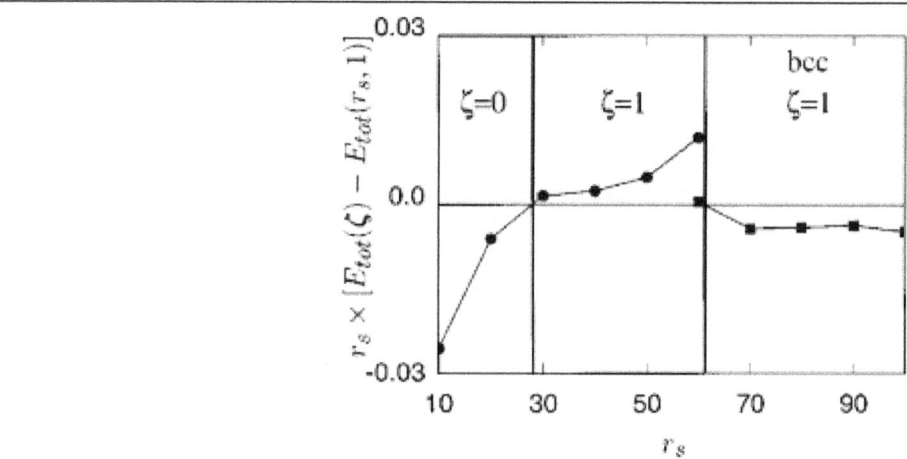

Total energy difference in eV vs r_s. Left to right: paramagnetic, ferromagnetic, ferromagnetic bcc crystal and the ferromagnetic gas.

Solid Hydrogen

The protons are localized around lattice sites → exchange effects between protons are negligible → the wave function can be taken to be symmetric:

$$\psi_T = \psi_e e^{-\sum_{\alpha\neq\beta}u_{pp}(d_\alpha - d_\beta) - \sum_\alpha c_\alpha |d_\alpha - \bar{d}_\alpha|^2}$$

- d_α – proton position,
- $\bar{d}_\alpha$ – lattice site,

---

[27] Ehrenfest classification: first-order phase transitions exhibit a discontinuity in the first derivative of the free energy with respect to some thermodynamic variable.

- ψ_e – electronic part of the wave function for fixed proton positions in a standard Slater-Jastrow form,
- u_{pp} – proton-proton correlation term
- Gaussian functions localize the protons around the lattice sites.

9.4 Excited states

VMC and DMC are designed for ground states, but they can be used for some excited states as well.

9.4.1 DMC for Solids

The lowest band gap of a solid may be measured with a photoemission and inverse photoemission experiments. The band gap energy can be obtained by the difference in energies in the N-electron system:

$$E_g = (E_{N+1} - E_N) - (E_N - E_{N-1})$$

Since the expression involves only ground-state energies → DMC can be used.

9.4.2 VMC many-electron model

Devise a many-electron ψ_T that models an excited state and use it in VMC. If the chosen ψ_T has a specific symmetry ∧ variational principle → $E \geq E_0$.

9.4.3 Fixed-node DMC

In DMC the wave function always evolves to the ground-state → DMC seams inapplicable to excited states. However, the nodal constraint ensures convergence to the lowest energy state compatible with the imposed nodal surface, not to the overall ground state. If the nodal surfaces of ψ_T and the exact wavefunction are the same → the fixed-node DMC gives the exact energy of that eigenstate.

9.5 Pseudopotentials

Problems:

i. Computational cost with increasing atomic number Z is in the interval $[Z^{5.5}, Z^{6.5}] \rightarrow$ impossible for heavier atoms.

ii. Core electrons move fast on smaller distances $\rightarrow$ there are a sharper gradients in the core region $\rightarrow$ requirement for the small time steps $\rightarrow$ reduce the efficiency.

iii. The energies of core electrons in heavy atoms are much greater than the energies of valence electrons $\rightarrow$ including valence electrons is very expensive when statistical error in the total energy must be reduced. In terms of local energy the core electrons are very noisy and they contribute a disproportionate share of the variance in local energies. In other words, fluctuations in local energy tend to be large near the nucleus because both kinetic and potential energies are large. The fluctuations can be reduced by a good ψ_T, but accurate wave functions are still not obtained and designed for the heavy atoms.

Observations:

i. The core remains almost independent of its environment.

ii. Many properties of interest (low-energy excitations, interatomic bonding and for chemistry in general) are determined by the valence electrons.

The solution is the use of pseudopotentials to remove the core electrons from the calculation (the loss in accuracy is very small), by reducing the effective value of $Z \rightarrow$ the computational cost's dependence on Z now is determined by the # of valence electrons that must be included in the simulation in order to obtain the satisfactory results.

Example: Si atom $[1s^2 2s^2 2p^6 3s^2 3p^2]$:

- $3s$ and $3p$ orbitals take part in the chemical bonding $\rightarrow$ they are labeled as valence orbitals,
- $1s, 2s$ and $2p$ orbitals stay with the atom $\rightarrow$ they are labeled as core orbitals.

We then create an effective potential (pseudopotential) that reproduces the effects of both the nucleus and the core electrons on the valence electrons.

The pseudopotential contains angular momentum projectors $\rightarrow$ it is a nonlocal operator.

$$V_I(R) = V_{\text{loc}}(R) + V_{\text{nl}} = \sum_i V_{\text{loc}}^{\text{ps}}(r_i) + \sum_i V_{\text{nl},i}^{\text{ps}}$$

- V_I – electron-ion dependent terms of ion I
- $V_{\text{nl},i}^{\text{ps}} = \sum_l \sum_{m=-l}^{l} V_{\text{nl},l}^{\text{ps}}(r_i) Y_{lm}(\Omega_i) \int Y_{lm}^*(\Omega_i') \, d\Omega_i'$ – short-ranged non-local potential, a correction for the angular momentum l, a nonlocal operator that acts on the arbitrary function of r_i, $f(r_i)$:

$$V_{\text{nl},i}^{\text{ps}} f(r_i) = \sum_l^{l} \sum_{m=-l}^{l} V_{\text{nl},l}^{\text{ps}}(r_i) Y_{lm}(\Omega_i) \int Y_{lm}^*(\Omega_i') f(r_i') d\Omega_i'$$

- V_{loc} – long-ranged local potential, local part common to all angular momenta. It can be directly evaluated during a MC simulation.

9.5.1 VMC

The action of the nonlocal pseudopotential on the wave function can be written as a sum of contributions from each electron and each angular momentum channel. The contribution to the $E_L = \frac{\mathcal{H}\Psi_T}{\Psi_T}$ made by the nonlocal pseudopotential terms (for simplicity we consider only a single atom placed at the origin):

$$V_{nl} = \frac{V_{nl}\Psi_T}{\Psi_T} = \sum_i \frac{V_{nl,i}^{ps}\Psi_T}{\Psi_T} = \sum_i V_{nl,i}$$

The nonlocal contribution of electron i to the E_L is:

$$V_{nl,i} = \sum_l V_{nl,l}^{ps}(r_i) \sum_{m=-l}^{l} Y_{lm}(\Omega_{r_i}) \int \frac{Y_{lm}^*\left(\Omega_{r_i'}\right)\Psi_T'}{\Psi_T} d\Omega_{r_i'}$$

- $\Psi_T = \Psi_T(r_1, \dots, r_{i-1}, r_i, r_{i+1}, \dots, r_N)$
- $\Psi_T' = \Psi_T(r_1, \dots, r_{i-1}, r_i', r_{i+1}, \dots, r_N)$

The integral is over the sphere passing through the ith electron and centered at the origin.

$$Y_l^m(\theta, \varphi) = (-1)^m \sqrt{\frac{(2l+1)}{4\pi}\frac{(l-m)!}{(l+m)!}} P_l^m(\cos\theta)e^{im\varphi}$$

We can simplify the equation for nonlocal potential by choosing the z-axis along $r_i \rightarrow Y_{lm}(0,0) = 0$ for $m \neq 0 \rightarrow Y_l^0 = \sqrt{\frac{2l+1}{4\pi}} \underbrace{P_l^0}_{P_l}(\cos\theta)$[28]:

$$V_{nl,i} = \sum_l V_{nl,l}^{ps}(r_i) \frac{2l+1}{4\pi} \int \frac{P_l(\cos\theta_i')\Psi_T'}{\Psi_T} d\Omega_{r_i'}$$

The integral must be evaluated numerically, since the projection operators depend on all inter-particle distances. The integrals are of low-dimensionality $\rightarrow$ grid based integration methods are more efficient than MC. In practice, 6-12 point grids give sufficient accuracy for first and second row elements, where the character of the wave function is dominated by s and p orbitals. 6 point grids are sufficient to integrate d momenta, and 12 point grids g momenta.

In principle the nonlocal energy should be summed over all the ionic cores and all electrons. However, since the nonlocal potential of each ion is short ranged, we need to sum only over the few atoms nearest to each electron. The additional angular integrations are computationally inexpensive within VMC because E_L is not required during the one-electron moves and do not need frequent evaluation. If E_L

9.5.2 DMC

We can apply the pseudopotential localization approximation to resolve the sign problem in DMC pseudopotentials, but the use of the nonlocal pseudopotentials is still problematic and expensive.

---

[28] For $m = 0$ associated Legendre polynomial is equal to the ordinary Legendre polynomial.

9.6 Supercell

Periodic systems containing a few hundreds of electrons are large enough to model most solid states. QMC simulation cells for perfect crystals usually consist of several primitive unit cells. The supercell method is the ubiquitous approach for the study of solid state periodic boundary condition systems. Periodic boundary conditions impose an artificial periodicity on interactions in the system. All interactions are periodic with the periodicity of the supercell.

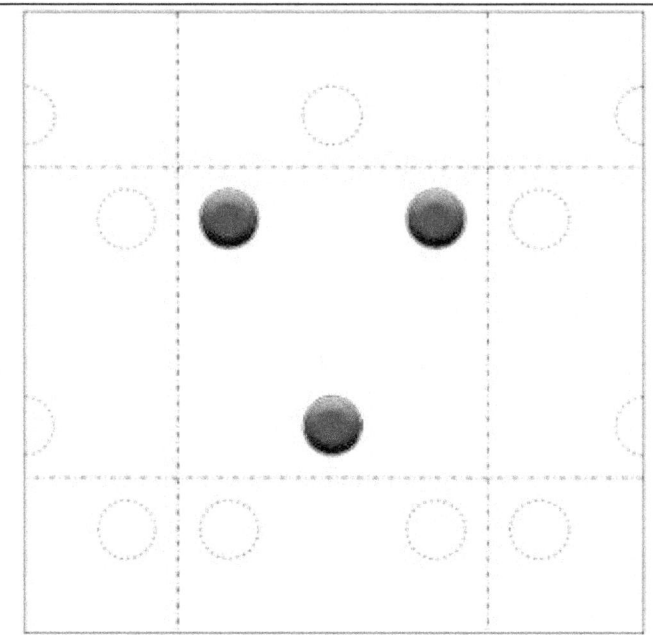

The supercell is in the center of the image. The "minimum image" must be applied when inter-particle distances are computed. The minimum distance can involve particle images in adjacent supercells.

Bloch's theorem may be applied to the wave functions: it allows the independent particle problem to be solved by considering only a single primitive unit cell. Bloch wave functions have the property that E_L has the full translational symmetry of the Hamiltonian. Bloch's theorem arises from the translational symmetry of the many-body Hamiltonian:

$$\mathcal{H} = -\frac{1}{2}\sum_{i=1}^{N}\nabla_i^2 + \sum_{i=1}^{N}V_{\text{ion}}(r_i) + \underbrace{\frac{1}{2}\sum_{R_s}\sum_{i=1}^{N}\sum_{j=1}^{N}\frac{1}{|r_{ij}-R_s|}}_{\text{interaction energy}}$$

(for $R_s = 0$ the self-interaction terms $i = j$ are excluded)

- $V_{\text{ion}}(r_i)$ – ionic potential – contributions from all the ions within the cell and their images,
- $\{R_s\}$ – the set of translation vectors of the cell lattice,
- N - # of electrons in the cell.

The invariance of the $\mathcal{H}$ under translation by R_s leads to many-body Bloch conditions (s – simulation-cell, p – primitive-cell):

$$\Psi_{k_s}(\{r_i\}) = U_{k_s}(\{r_i\})e^{ik_s\sum_{i=1}^{N}r_i}, \Psi_{k_p}(\{r_i\}) = W_{k_s}(\{r_i\})e^{ik_p\frac{1}{N}\sum_{i=1}^{N}r_i}$$

- U_{k_s} – the simulation-cell periodic part, antisymmetric under particle exchange,
- W_{k_p} – the invariant under the simultaneous translation of all electrons by any primitive lattice vector R_p, antisymmetric under particle exchange,
- k_s – simulation-cell wave vector, may always be reduced into the first Brillouin zone of the simulation-cell reciprocal lattice,
- k_p – primitive wave vector, may be reduced into the first Brillouin zone of the primitive reciprocal lattice, gives the physical crystal momentum measured in experiments.

The potential energy of charges q_i (electrons and nuclei) at positions r_i can be written as:

$$V = \frac{1}{2} \sum_i \sum_{j \neq i} \frac{q_i q_j}{r_{ij}} = \frac{1}{2} \sum_i q_i V_i$$

$V_i = \sum_{R_s} \sum_{j=1}^{N} \frac{q_j}{|r_{ij}+R_s|}$ – potential at r_i due to every charge except q_i. In QMC it depends only on the positions of the at most few hundred charges in the simulation-cell and is a poor imitation of the potential of the real solid, which depends on the positions of $\sim N_A$ electrons and ions.

The wave function can be chosen to satisfy the simulation-cell and primitive-cell Bloch conditions simultaneously – the wave functions must be identical if we translate all electrons by R_s.

The sum in the interaction energy is divergent, a problem resolved by the Ewald method: the point charge distribution is modified to give 2 convergent series which sum gives the correct value for the unmodified distribution. The neutral Ewald charge density is given by a sum over the point charges plus a screening background:

$$\rho_E(r) = \sum_{R_s} \delta(r - r_i - R_s) - \rho_{\text{background}}$$

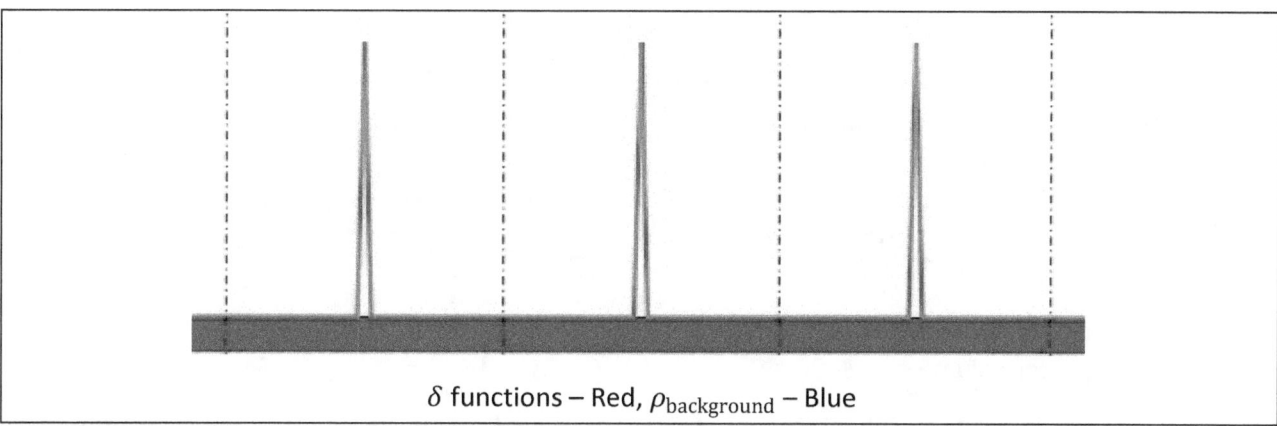

δ functions – Red, $\rho_{\text{background}}$ – Blue

The Ewald potential is a sum of two components (μ – width of the Gaussians):

1. the array of Gaussians which cancel the background charge density:

$$\rho_1(r) = \frac{1}{\mu\sqrt{\pi}} \sum_{R_s} e^{-\frac{(r-r_i-R_s)^2}{\mu^2}} - \rho_{\text{background}}$$

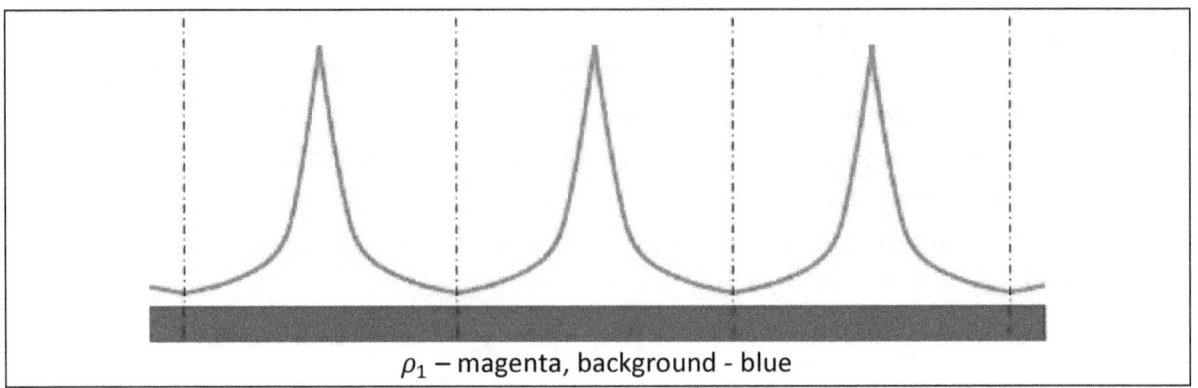

ρ_1 – magenta, background - blue

2. the array of Gaussians which are subtracted from the delta function charge distribution:

$$\rho_2(r) = \sum_{R_s} \left(\delta(r - r_i - R_s) - \frac{1}{\mu\sqrt{\pi}} e^{-\frac{(r-r_i-R_s)^2}{\mu^2}} \right)$$

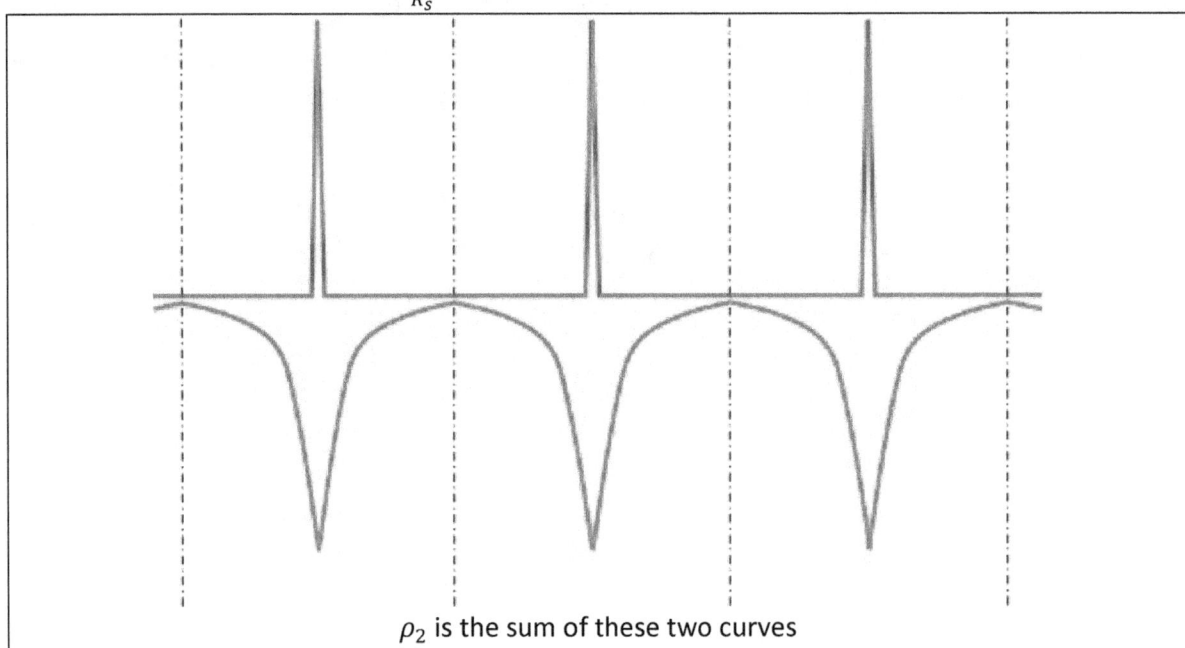

ρ_2 is the sum of these two curves

In reciprocal space, overall Ewald potential is:

$$V_E(r) = \sum_{G \neq 0} \frac{4\pi}{VG^2} e^{-\frac{\mu^2 G^2}{4} + i(r-r_i)G} + \sum_{R_s} \frac{\text{erfc}\left(\frac{|r - r_i - R_s|}{\mu}\right)}{|r - r_i - R_s|} - \frac{\pi\mu^2}{V}$$

- G – supercell reciprocal lattice vectors,
- V 0 supercell volume.

V_E can be used to compute the electron-ion potential; the Ewald sum for the electron-electron potential is similar, but the "self-image"[29] of an electron is removed. Electrostatic energies are computed by multiplying V_E by $q_i q_j$.

---

[29] when $i = j$.

9.7 The cost, the error and the parallelization

9.7.1 The error

Accurate trial wave functions are necessary to achieve high accuracy at a reasonable computational cost, but since the DMC wave function is generated stochastically[30], the results are largely free of the errors caused by the limited basis set used in most other techniques. By contrast, the quality of results obtained using the less accurate VMC is entirely determined by the quality of the ψ_T.

For small systems the DMC is capable of reaching chemical accuracy.[31]

9.7.2 The cost

The computational cost of a QMC calculation increases as the cube of the number of electrons: N^3 → the computer time required to calculate the energy of a system to some given accuracy using the fermion VMC and DMC scales as N^3 → applications to large systems are feasible.

The answers are obtained with a statistical error bar that decays only as the inverse of the square root of the computer time → the total amount of time required for accurate QMC calculations is quite large.

9.7.3 Parallelization

Algorithms are simple to program and are well suited to parallel computation. Parallel computation greatly reduces the time needed for completion.

VMC

To parallelize the VMC algorithm it is sufficient that each calculation uses a different random number sequence, then performing independent VMC calculations on each node of a parallel machine. Communication between the nodes is only necessary to obtain global averages (thus it is negligible).

DMC

In the parallel DMC algorithm, walkers are initially evenly distributed over all nodes, and each node is responsible for propagating its walkers. After each ensemble move, the walkers are redistributed as evenly as possible across all of the nodes.

---

[30] A stochastic process, or often random process, is a collection of random variables representing the evolution of some system of random values over time (for example, Markov chain).

[31] $1\frac{kcal}{mol} \approx .04\frac{eV}{molecule}$

130

Bibliography

1. William H. Press et al. *Numerical Recipes*, New York: Cambridge University Press, 2007.
2. William H. Press et al. *Numerical Recipes Webnote No.9, Complete VEGAS Code*: 2007. <http://numerical.recipes/webnotes/nr3web9.pdf>
3. Paul Richard, Charles Kent, *Techniques and Applications of Quantum Monte Carlo*: University of Cambridge, 1999.
4. John Shumway, David Ceperley, *Quantum Monte Carlo methods in study of Nanostructures*: 2004. <https://www.researchgate.net/publication/225067720_Quantum_Monte_Carlo_Methods_in_the_Study_of_Nanostructures>
5. Bernd A. Berg, *Markov Chain Monte Carlo Simulations and Their Statistical Analysis*, Florida State University: World Scientific Publishing, 2004.
6. John Tsitsiklis. *Probabilistic Systems Analysis and Applied Probability, Fall 2010*. (Massachusetts Institute of Technology: MIT OpenCourseWare)
7. David B. Thomas, Wayne Luk. *Gaussian Random Number Generators*, ACM Comput. Surv. 39, 4, Article 11, 2007. <http://doi.acm.org/10.1145/1287620.1287622>
8. Kari Rummukainen , *Monte Carlo Simulations in Physics*, University of Oulu. <http://www.helsinki.fi/~rummukai/lectures/montecarlo_oulu/lectures/mc_notes1.pdf>
9. Jonas Golde, *Monte Carlo integration: Adaptive Methods (VEGAS)*: Technische Universitat Dresden, 2013.
10. Jessy Liberty, Bradley L. Jones, *Teach Yourself C++ in 21 Days*: Sams Publishing, 2004.
11. B. P. Demidovich, I. A. Maron, *Computational Mathematics*, Moscow: Mir Publishers, 1981.
12. B. V. Gnedenko, *The Theory of Probability*, Moscow: Mir Publishers, 1978.
13. James B. Anderson, *Diffusion and Green's Function Quantum Monte Carlo Methods*: Published in J. Grotendorst at al. *Quantum Simulations of Complex Many-Body Systems: From Theory to Algorithms*, Julich: John von Neumann Institute for Computing, 2002.
14. Ben Schofield, *Fixed-Node Diffusion Monte Carlo*.
15. I. Kosztin at al. *Introduction to the Diffusion Monte Carlo Method*, Illinois: University of Illinois, 1995.
16. Introduction to the variational and diffusion Monte Carlo methods, 2015., J.Toulouse et al.
17. W.M.C. Faulkes at al. *Quantum Monte Carlo simulations of solids*, Cambridge: TCM Group, Cavendish Laboratory, 2001.
18. B. Abdullaev at al. *Complex Diffusion Monte-Carlo method for the systems with complex wave function: test by the simulation of 2D electron in uniform magnetic field*, Tashkent: Tashkent State University, 2013.
19. M. Hjort-Jensen, *Computational physics*, Oslo: University of Oslo, 2003.

www.ingramcontent.com/pod-product-compliance
Lightning Source LLC
Chambersburg PA
CBHW080659190526
45169CB00006B/2187